The President's Salmon

The President's Salmon

*Restoring the King of Fish and
Its Home Waters*

Catherine Schmitt

Down East Books

Camden, Maine

Published by Down East Books
A wholly owned subsidiary of The Rowman & Littlefield Publishing Group, Inc.
4501 Forbes Boulevard, Suite 200, Lanham, Maryland 20706
www.rowman.com

Unit A, Whitacre Mews, 26-34 Stannary Street, London SE11 4AB

Distributed by National Book Network
Copyright © 2015 by Catherine Schmitt

All rights reserved. No part of this book may be reproduced in any form or by any electronic or mechanical means, including information storage and retrieval systems, without written permission from the publisher, except by a reviewer who may quote passages in a review.

British Library Cataloguing in Publication Information Available

Library of Congress Cataloging-in-Publication Data

Schmitt, Catherine, 1976-
The president's salmon : restoring the king of fish and its home waters / Catherine Schmitt.
pages cm
Includes bibliographical references and index.
ISBN 978-1-60893-408-9 (cloth : alk. paper) -- ISBN 978-1-60893-410-2 (electronic)
1. Atlantic salmon--History. 2. Atlantic salmon fisheries--Maine--Penobscot River--History. 3. Rare fishes--Maine--Penobscot River--History. I. Title.
QL638.S2S278 2015
597.5'6--dc23
2015009024

∞™ The paper used in this publication meets the minimum requirements of American National Standard for Information Sciences Permanence of Paper for Printed Library Materials, ANSI/NISO Z39.48-1992.

Printed in the United States of America

Contents

Acknowledgements — vii

1. The President's Salmon — 1
2. Seboomook — 9
3. A Tradition Begins — 19
4. Chesuncook — 33
5. Big Business — 47
6. Wassataquoik — 63
7. Empty Nets — 79
8. Penobscot — 93
9. Troubled Waters — 105
10. Kenduskeag — 115
11. Welcome to Salmon City — 127
12. Alamoosook — 141
13. Paper Salmon — 151
14. Wassumkeag — 161

Notes — 173

Selected Bibliography — 215

Index — 221

Acknowledgements

I am grateful to everyone who helped me over the years as this book evolved.

I started writing about the Penobscot River while working as a research assistant at the Senator George J. Mitchell Center at the University of Maine, with funding from the U.S. Geological Survey Maine Water Resources Research Institute and an Olin Fellowship from the Atlantic Salmon Federation. I benefited greatly from my work with scientists and others studying the restoration of the Penobscot River, supported by the Maine Department of Environmental Protection, Maine Sea Grant, and the Penobscot River Restoration Trust.

Some of the content appeared in different form in the *Bangor Daily News, Northern Sky News, The Working Waterfront, Atlantic Salmon Journal,* and *Maine Boats, Homes & Harbors*, under the editorship of Murray Carpenter, David Platt, Tom Groening, Martin Silverstone, and Gretchen Ogden.

I conducted much research in the University of Maine Fogler Library, in particular Special Collections, with generous assistance from Desiree Butterfield-Nagy, Elaine Smith, and Bonnie Garceau, and the staff in the Circulation and Reference departments. The Bangor Room and newspaper collections of the Bangor Public Library were invaluable, as was the Maine State Archives and Curtis Bentley at the Maine Law & Legislative Library. Gretchen Faulkner of the Hudson Museum and Julia Clark of the Abbe Museum helped me track down and view a Wabanaki salmon spear. Additional research and writing was conducted in Belfast Free Library, Bowdoin College Library, Glickman Library, Hutchinson Center, Portland Public Library, and Skidompha Library.

Many people took the time to show me the river and its fish, and to share their knowledge with me: John Banks, James E. Francis, Jr., Dan Kusnierz, Dan McCaw, and the late Clem Fay, of the Penobscot Nation Department of Natural Resources; Randy Spencer, Ernie Atkinson, Joan Trial of the Maine Department of Marine Resources Bureau of Sea Run Fisheries and Habitat and their predecessors at the Atlantic Salmon Commission, Ed Baum and Al Meister; Rory Saunders, John Kocik, Dan Kircheis, Tim Sheehan, and Justin Stevens of the National Marine Fisheries Service; Chris Domina, Peter Steenstra, Peter Lamothe, and staff of the U.S. Fish and Wildlife Service. Also Senator George J. Mitchell, David Courtemanch, Dean Cutting, Adria Elskus, Richard Judd, Steve Kahl, Michael Kinnison, James McCleave, Sandra Neily, Alice Kelley, Brian Robinson, Laura Rose Day, Josh Royte, Bill Townsend, Gayle Zydlewski, and Joe Zydlewski. Gunnar Knapp, Laura Lindenfeld, and Sandra Oliver informed my study of the history of salmon as food. Evan Johnston in the Office of Representative Chellie Pingree secured my entry to the White House. Thanks to Tara Trinko-Lake for creating the watershed map.

I have tried to honor the members—and memories—of the Penobscot, Eddington, and Veazie Salmon Clubs, especially Royce Day, Roger D'Errico, Don Foster, Lou Horvath, the late Dick Ruhlin, and Claude Westfall.

Many of these people also reviewed portions of the text for scientific and historic accuracy. Thanks to Michael Steere for his edits and support for the book; any remaining errors are mine.

I am fortunate to have the support of Paul Anderson, co-workers at Maine Sea Grant, and colleagues at the University of Maine. Gratitude extends to my Stonecoast mentors Barbara Hurd, Debra Marquart, Rick Bass, and especially Jaed Coffin, and all the workshop participants who gave feedback on earlier drafts. A huge thanks to my "readers" Marybeth McGinley and Chandra McGee for their thoughtful reactions to the final draft, also to Linda Buckmaster, Jennifer Bunting, Kathleen Ellis, and Ted Williams for their encouragement. Thanks to Bowen's Tavern, The Brick House, and Schooner Landing for allowing me space to work.

To my family and friends, thank you for understanding my absence when writing couldn't wait, and to E.G., the most understanding of all.

Finally, to everyone, past, present, and future, who has worked to restore the Penobscot River, thank you. This is your story.

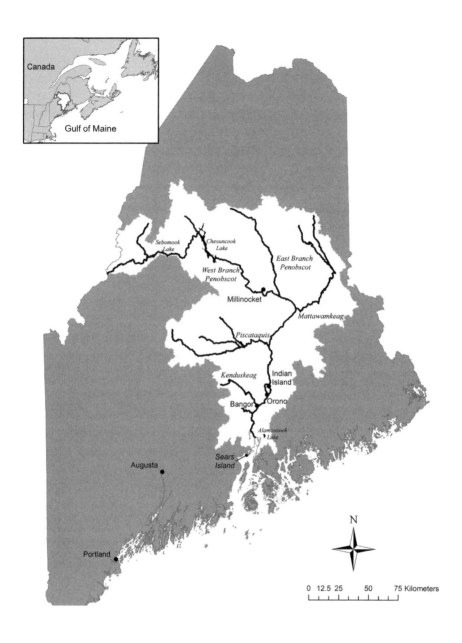

Chapter One

The President's Salmon

Monday, May 25, 1992

Claude and Rosemae Westfall drove their Buick south on Maine's I-95. Claude was dressed sharply if atypically in a green suit jacket and white button-down shirt; Rosemae wore her best black dress and a pink blazer. In the backseat, a styrofoam cooler of ice held an Atlantic salmon Claude had caught in the Penobscot River three weeks earlier.

On their way south, the Westfalls crossed the Kennebec, Androscoggin, and Saco rivers, all of which once had thriving populations of salmon.

Every spring, for thousands of years, the rivers that empty into the North Atlantic Ocean turned silver with migrating fish. Tomcod and rainbow smelt swam out from beneath melting ice. Young eels found their way from the Sargasso Sea. Alewives and blueback herring surged upstream when serviceberry trees bloomed white in still-bare woods, followed by shad in June, then sea lamprey, striped bass, and sturgeon.

Among the crowded schools swam the king of fish, the Atlantic salmon. From New York to Labrador, from Russia to Portugal, sea-bright salmon defied current, tide, and gravity, driven inland by instinct and memory to the very streams where they themselves emerged from gravel nests years before.

They were fish worthy of story, myth, and legend; in North America, they provided sustenance to many and supported major commercial fisheries.

Hundreds of thousands of salmon used to ascend the rivers of New England. By 1992, no adult salmon returned to the Kennebec River. Seventeen came back to the Androscoggin, and only eight to the Saco. Their banks were empty of salmon anglers.

On the Penobscot, the largest river in Maine and the second-largest in New England, Claude's fish was the first of 2,386 salmon that would return to the river that year, more than 70 percent of the total returns along the entire East Coast of the United States, but only 2 percent of the river's historic population.

After two and a half hours on the highway, Claude took the Route 1 exit and followed the signs to Kennebunkport. Looking out the window, Rosemae marveled at the mansions lining the two-lane highway: eighteenth-century colonials, nineteenth-century Victorians, Greek Revivals, the "Wedding Cake House." Claude drove on through the village of antique shops, art galleries, restaurants, and souvenir stands just opening up for the summer tourist season. Rosemae remarked on the colorful striped awnings and freshly painted signs. White-fenced inns and yacht clubs overlooked sailboats in the harbor. Kennebunkport looked nothing like the college town of Orono, where Claude and Rosemae lived, on the banks of the Penobscot River.

Claude crossed the sparkling Kennebunk River and turned right on Ocean Avenue. Soon they came to a roadblock. Claude stopped.

"Where are you going?" asked a security guard.

"I have an appointment," said Claude.

The guard scanned his clipboard.

"Oh, yes, Mr. Westfall, go right ahead."

Claude drove on, and from an overlook their destination came into view: a large gray-shingled home, buildings, tennis courts, lawn, all neatly arranged on a point jutting into the sea. Eider ducks bobbed on the waves between dark, jagged rocks. Claude turned right at the entrance, and stopped before a tall, iron gate flanked by stone pillars crusted orange with sunburst lichen.

More guards examined the bottom of the car, looked under the hood, glanced at the fish in the cooler in the backseat, and inspected the other gifts. There was a blue cap from the Veazie Salmon Club—one of the three clubs on the Penobscot that served as social networks for sport fishermen, provided access to the best fishing pools, and advocated for the restoration of Atlantic salmon. A carved wooden box held hand-tied salmon flies, including one Claude had tied himself: wings of peacock sword feathers and fur from a squirrel's tail, dyed a fluorescent green, wrapped on a lapel pin with silver, black, and green floss—the CZ Special, the same fly he used to catch the fish that was in the cooler. Claude's namesake creation was one of many fly patterns he had developed in nearly forty years of fishing for Atlantic salmon.

Finding everything clear, the guards waved the Westfalls through to the heliport, where they parked and were escorted to an outbuilding nearby. A guard carried the cooler and the other gifts to an office near the landing pad. The guard opened the cooler. The ice had begun to melt. The fish looked more like something behind glass at the seafood counter than nine and a half pounds of flesh and bone that had muscled its way from river to sea and back again. Black spots peppered its silver back; its sharply defined dorsal fin revealed it to be a wild fish, not one that had begun life in a hatchery. Hatcheries had been supplementing Maine's salmon population since the late 1800s.

The guard brought the cooler and gifts out to the lawn as President George H. W. Bush, Barbara Bush, Maine Representative Olympia Snowe, and John McKernan, Snowe's husband and the governor of Maine, came down from the main house to join the Westfalls and Maine Commissioner of Marine Resources Bill Brennan in front of the cedar-shingled building. Millie, the Bush family's springer spaniel, sniffed around the cooler. Claude gave the president the blue Veazie Salmon Club cap. At the photographer's cue, Claude held up the slightly defrosted fish and handed it to the president, who grasped the fish by the line that had been looped around its tail.

The small crowd giggled as the fish dangled. Claude said to the president, "You know, we have several things in common."

"What's that?" asked the president.

"My mother's great grandfather was Henderson P. Bush, who fought in the Civil War. It's the same Bush family."

"Well," said President Bush, "That makes us kissing cousins, doesn't it?"

"The other thing we have in common," said Claude, "is that I'm a Republican." Bush nodded, smiled, and looked at Republican Governor McKernan.

"Governor, you'll be glad to hear this, too."

Everyone laughed as the security guard took the fish and put it back in the cooler.

President Bush turned to Claude. "Mr. and Mrs. Westfall, we have a little time left. Would you like to come up to the house?"

Thrilled, the Westfalls accepted the invitation.

The security guard took the cooler around to the service entrance of the main house. The party, trailed by Secret Service agents, walked along the shoreline path lined with red tulips, salt-spray rose, and security cameras. Out on the Atlantic Ocean, two Coast Guard cutters patrolled the waves.

In the entranceway of the house, Claude recognized people in the framed photos hanging on the wall: Ronald Reagan, French president Francois Mitterrand, tennis player Chris Evert, Bush himself fishing off the Maine coast.

Bush joined a long line of fishing presidents that began with George Washington, who made a profit netting sturgeon, herring, and shad from the Potomac at Mount Vernon. Chester Arthur once caught a world record fifty-one-pound Atlantic salmon in the Cascapedia River in Quebec. His successor, Grover Cleveland, took fly rods on his honeymoon.

Although Bush had been salmon fishing in Canada, he did not claim to be a skilled fly fisherman. "I've fished a couple of Oregon rivers for steelhead, without much luck," Bush told *Fly Rod & Reel* magazine in 1989. He was a troller, a spinner, a bait-caster, racing to ocean fishing grounds in his cigarette boat, *Fidelity*. He caught bonefish, tarpon, sailfish, mackerel, barracuda, stripers, but rarely ate them. Bush didn't much care for seafood, much less where it came from. He usually threw back his catch or gave it to Secret Service agents before going inside to chow down on beef jerky, nachos, chili, or barbecued ribs.

The Westfalls and the others sat in the living room. President Bush asked the kitchen to bring out something to eat. Through the swinging doors into the hallway, Claude could see the styrofoam cooler that held the salmon.

The human relationship with salmon has always been tinged with mystery and myth. The oldest record of this relationship dates back 40,000 years, when humans left salmon bones in a cave in what is now Spain. And 22,000 years ago, someone carved a life-size image of an Atlantic salmon into the floor of a cave in southern France. Six to ten thousand years ago, people painted and etched salmon into the rocks along the northern shore of Norway and Sweden.

The salmon's effortless leaping and ability to survive in both river and sea led the Celts to mythologize the salmon as holder of all mysterious knowledge, gained by consuming the nine hazelnuts of wisdom that fell into the

Well of Segais. The salmon is said to be as old as time and to know all the past and future. Knowledge is gained with each leap against the current, each finger burned by fire-roasted flesh; each salmon devoured, its bones thrown back to the river.

Salmon appear in the Arthurian legends as the all-knowing water creature. Heroes ride the back of the great salmon in search of missing sun gods. Roman conquerers counted salmon among their favorite foods. In the first century, Pliny the Elder wrote that in the Aquitania region of southern France, "the river salmon is preferred to all the fish that swim in the sea."

In the eleventh century, the Vikings journeyed to Vinland (North America) and inscribed the presence of salmon in the runic saga of Eric the Red. The native people of Northeastern North America favored salmon, too, and they speared great numbers of the large, flavorful fish. Smoked salmon could feed a family through the long winter. The indigenous Beothuks of Newfoundland buried baskets of dried or smoked salmon with the dead, and those along the Restigouche River reproduced images of salmon on their clothing and canoes.

The salmon is associated with prophecy and inspiration because of its instinct to find its distant homeplace, and its ability to reconnect those in the present with ancestral knowledge, and to put it to practical use. As a food, salmon is high in calories and nutritious.

Fresh salmon appeared at the feasts of rulers and kings. The reddish flesh of salmon and its rich flavor meant that the fish was sometimes regarded as "red blooded" and was included on the list of "royal" fish alongside sturgeon and whales. Is this why Isaak Walton declared, "The Salmon is accounted the King of freshwater-Fish, and is ever bred in Rivers relating to the Sea"?

Since 1912, salmon anglers had been giving the first salmon caught in the Penobscot River each year to the president of the United States. But by the early 1990s, the presentation of the presidential salmon was a token, an opportunity for political posturing, a symbol of both the tenuous status of the species and the promise of renewal that had infused the Penobscot River for as long as the river had been abused.

In the spring of 1992, Haitian refugees were washing ashore in Florida. Violent conflict simmered in Croatia. Developing nations prepared to band together against imperialism at the upcoming Earth Summit in Rio. The Persian Gulf War had ended, but oil fires still burned in Kuwait and Saddam Hussein remained in power in Iraq.

At home, the economy staggered beneath the weight of a then-record $300 billion federal deficit. Several years of high energy prices had prompted new federal legislation to deregulate electricity producers and allow them access to the nation's power grid, including hydroelectric dams. Tax breaks and new laws that required utilities to purchase power from renewable sources had infected developers with dam fever.

In Maine, Claude Westfall and his fellow salmon club members had just defeated a proposal to reconstruct an old dam on the Penobscot River in Bangor, and now they faced yet another dam proposal upstream in nearby Orono.

A veteran and recently retired engineering professor, Claude knew it was not his place to discuss the dam proposals with President Bush. Nor did he bring up the fact that the year before, the U.S. Fish and Wildlife Service had designated Atlantic salmon in five rivers east of the Penobscot as candidates for listing under the Endangered Species Act.

In his 1988 campaign, Bush had promised to make amends for the Reagan administration's crimes against nature and to be the Environmental President. While he advanced policies to reduce acid rain and improve water quality, Bush advocated a National Energy Strategy that included more hydroelectric development, and reduced the ability of federal wildlife officials to intervene, making it easier for developers to obtain licenses. At the same time, his administration was taking steps to weaken the Endangered Species Act.

Just one week before receiving Claude's salmon, Bush had waived protections for the Northern spotted owl to allow logging in the Pacific Northwest. He would soon proclaim, "The Endangered Species Act was intended as a shield for species against the effects of major construction projects . . . not a sword aimed at the jobs, families, and communities of entire regions. . . . It's time to put people ahead of owls." These words echoed across the country and throughout the Penobscot River watershed, where logging giant Great Northern Paper owned 10 percent of all the land in Maine, employed thousands of people in the woods and mills, and operated the nation's largest private hydroelectric system on waters home to dwindling numbers of Atlantic salmon.

Bush, who was known to take salmon fishing vacations in Canada, was about to make opposition to the Endangered Species Act a centerpiece of his 1992 campaign. He would call the Act a broken law that would not stand. When the Act threatened to block a dam project in Bush's home state of

Texas, Interior Secretary Manuel Lujan urged that the law be weakened. Lujan, a man who said he couldn't tell the difference between a red squirrel and a black squirrel, was not shy about his attitude toward the "too tough" law.

Lujan delayed designating critical habitat for the threatened spotted owl, and instead planned to allow the owl to become extinct in areas of Washington State around Olympic and North Cascades National Parks. Owls would be captured and moved south to other federal lands so that logging could continue around the parks. On the Penobscot, industry put forth a similar plan that would trap salmon and truck them upstream so that dam construction could move forward.

The Westfalls ate donuts and chatted with the Bushes and the other officials for about an hour before leaving for the long drive back to Orono. On the way home, Claude and Rosemae discussed their time with the president.

"I think we really hit it off," said Claude, who couldn't believe the whole thing was over. It all went by so fast.

"You are close in age," said Rosemae. Claude was sixty-three, the president sixty-seven.

"I wasn't nervous at all," said Claude, thinking about the ease of the conversation. He did wonder about the salmon, if the president would eat it, what would happen to it.

"I had a great time," said Rosemae.

"Me too," said Claude.

Upon returning home to Orono, Claude wrote President Bush a letter, thanking him for the visit and inviting him to go salmon fishing on the Penobscot. The president politely declined, "due to other pressing commitments."

Claude Westfall had no idea he would be the last fisherman to present the first wild Atlantic salmon caught in Maine's Penobscot River to the president of the United States. In the years that followed, the fish, the eighty-year tradition of a gift, became disconnected from its landscape, its people, and the laws that were meant to protect them both.

Chapter Two

Seboomook

The Penobscot River watershed covers more than eight thousand square miles, almost one third of Maine. The river falls thousands of feet. Along the way, it collects rain shed by great mountains, roars crystal clear down gorges and waterfalls, flows around boulders and ledges, and meanders through forested hills and lowlands. Groundwater seeps into the channel as it trickles slow and dark through extensive bogs and swamps, braids around islands and rapids, and ebbs and floods in tidewater and marsh.

Nearly the entire watershed is salmon country, and all of it was once covered in a mile-thick sheet of ice. These facts are not coincidental. The recent and not so recent physical history of the Penobscot River watershed shaped the landscape, the flora and fauna, and the patterns of human activity that led to the Penobscot River of today.

The deeper history of the landscape, the one that led to the rocks and mountains the ice worked upon, involves hundreds of millions of years of expanding and shrinking oceans; exploding volcanoes; colliding tectonic plates; crumpled, twisted, and crystalized layers of sediment; ocean floor thrust skyward; erosion and weathering. Then came two million years of ice ages. At the end of the most recent glaciation, when the Laurentide ice sheet covered northeastern North America, the Penobscot River emerged and the first salmon entered the landscape that was not yet Maine.

The ice extended far out into the North Atlantic, calving icebergs into the sea. Advancing ice scraped clean the hard granitic and volcanic rocks of Katahdin and the surrounding mountain landscape, and eroded low areas of fine-grained sedimentary bedrock into rolling hills and lowlands. About 14,000 years ago, the climate warmed and the glacier began to melt.

At Seboomook Lake, more than two hundred miles from the ocean, the North and South Branches of the Penobscot River come together to form the West Branch Penobscot River. Today, the lake is part of recreational lands owned by the state and surrounded by industrial forestland. Seboomook is a place where history is close to the surface, where it is possible to imagine another time, when the last of the ice was melting away, and salmon were among the first to arrive.

Salmonids are among the most northern of freshwater fish. Along with their relative, the Arctic char, Atlantic salmon are one of the first to inhabit an area when a glacier leaves. They don't need trees, shrubs, or dirt, only rough, cold-filtered gravel to shelter their eggs. They don't need food for themselves, only enough insects in the overlying water for their offspring to eat. Salmon spawn in fresh, stony gravel created by glacial runoff; at times within miles of the ice's edge. Their migratory tendency and marine tolerance meant that they could expand to more distant terrain than fish restricted to fresh water. Thus salmon were able to take advantage of new habitat when the glacier receded.

When the ice shrank northward, the sea followed, flooding low-lying areas depressed from the weight of so much ice, extending up the Penobscot Valley as far as Medway, creating islands of the coastal mountaintops. For a few thousand years, ocean waves crashed against hillsides. Marine silt and clay settled to the bottom of the flooded areas, capturing shells of mussels, clams, and barnacles. Walrus and whales swam inland. But the land had already begun to rebound and rise above the sea.

Salmon that had traveled into the deeper basins—what became Sebec, Green, West Grand, and Sebago lakes—became stranded, landlocked, but they still carried in their cells a prehistoric ability to survive in freshwater. They adapted to lake food sources, and migrated short distances into flowages and feeder streams to spawn.

The rest of the salmon continued to migrate through the drainage system of the emerging Penobscot, which surged with glacial runoff, raging down to the distant sea. Deranged by thaw, the river reoccupied older valleys from previous interglacial periods; where the glacier had blocked old channels with sediment, the river found new routes to the sea over bare ledges and boulders.

On the exposed plain, punctuated by gravel ridges and bedrock hills, winds sculpted sand into dunes. People moved in from the south and west, camping between the dunes, hunting caribou. New grass covered the sand dunes; tundra of stunted willow and birch sprouted across the hilltops. People followed the watery trails of melted ice.

White pine spread into open woodlands of poplar, spruce, and larch; then oak and other hardwoods appeared about 8,400 years ago. Hemlock and pine alternately dominated for thousands of years after that, then birch and beech moved in.

Where the sea had left behind a thick layer of impermeable blue-gray mud, basins filled with rain, became expanses of open water and then cattail marshes. Into the marshes came fish, birds, turtles, muskrat, and beaver. The people walked along the exposed sand and gravel eskers, footsteps across the rim of an ancient lake.

When the lightning bugs returned in early summer, the people netted salmon from the streams. From metamorphic rock and pearly shale, they chipped, pecked, and ground stone into tools, rods, and other objects. In the river banks, they found deposits of *olamon*, powdered hematite, red ochre to mark their path, to bury their dead with ceremony and honor at sandy banks, pond outlets, and high points overlooking the river, the special places where they would gather during fishing season. In the graves they placed finely crafted plummets, pear-shaped stones for weighting fishing nets, so the dead could continue to fish in the afterlife.

On the shore of Seboomook Lake, atop a sandy bluff between knolls covered in white pine, fragments of burned turtle shells are preserved in the shallow, acidic soils beneath hemlock and pine. Three thousand years ago, native canoeists camped here in summer, maybe on their way to and from Munsungan Lake to collect shiny red chert. Along the high cliffs of Seboomook, stone shapers gathered at a small workshop to craft tools. They cooked fish and turtles pulled from the river.

Archaeology is based on fragmentary evidence and interpretations can be speculative. Buried cultural history is subject to degradation, erosion, and warping as the landscape evolves, making establishing chronology difficult. Native people lived and moved through the entire region, but archaeologists can only see places where they stopped to make stone tools, or cook food over a fire. The Penobscot River flows through hundreds of occupation sites. Most date from the last six thousand years, but some are older. Two spear points tell that people were in the Penobscot Valley 10,000 years ago, using the river as a travel route.

Maine's acidic soils tend to dissolve skin, fur, and bone. Organic material only survives in coastal shell middens, or where it was burned in very hot fires. The burned bones become calcified and brittle, fragmenting into smaller pieces. The oldest salmon bones that have been found date to 7,000 years ago.

The story of this river is in pieces: bits of research based on scraps of history dug from lake shores, river banks, and coastal islands. Charred chips of salmon bone from streamside hearths. And fragments of memory, passed down through generations of people.

Fragments tell that humans first entered the landscape of Maine at the end of the last Ice Age, 14,000-12,000 years ago, when nomadic people migrated into the landscape from areas that had not been under ice. The modern Penobscots (*Burnurwurbskek*), who, along with the Passamaquoddy, Maliseet, and Micmacs, are members of the Wabanaki Confederacy of Native American Tribes, trace their ancestry to the beginning of time. They have always been here.

The Wabanaki, People of the Dawn, came from the east, floating across the ocean on an island; they came from the north, south, and west, toward the sun; they knew the walrus and the mammoth, the pileated woodpecker and the first salmon.

All things move. Great white swans came from the east. The Christians, in search of the gold-drenched land of Norumbega, asked for directions, asked the river people, *Where did you come from?* The natives pointed to the river, to the ash where it split into strips, to the birch where the bark peeled away, to the islands upstream where fiddlehead ferns unfurled each spring, and to the marshes downstream where the salt air was heavy with the vanilla scent of sweetgrass.

The strangers followed or were guided by the natives along the river and the trails worn between the waters where birchbark canoes were carried from cove to stream to lake. The strangers borrowed local words, like *Penobscot,* to describe the river—place where rocks widen out, the strange new place. This is how the language worked, from the perspective of a canoe moving upstream, naming currents, rapids, the river bottom and ledges, thickly forested banks and the flattest places to cross: *ahwangan*, a portage between streams.

Where did you come from?

The natives pointed to the carries and the streams, always to the water. The strangers were like a great fog cast over the world. The strangers took to the water, catching fish and hunting beaver. They hunted beaver until the beaver were gone and the beaver dams fell apart and everything drained, dried up until in some places a person could no longer paddle from one place to the other. The native people began to starve.

Where did you come from?

Land, it was always about land, and the native was an enemy, a traitor who had what the strangers wanted—land, water, wealth—and so they put a bounty on his head and the scalps of his children. The leaders of the new nation presented the natives with a piece of paper to sign that would determine where they would be from—always papers to sign, and so they signed a treaty retaining the islands in the river. New settlements continued to appear. Neither foreign proclamations nor the new American president, George Washington, recognized the sovereignty of the native people.

Paper was a trick, a trap, the tool of thieves.

The Penobscots were signing treaties when they should have been out fishing, because the fish were running and fish were food. The falls at Old Town and the twelve islands in the currents below were the prime place for fishing, the spring place where hundreds of families reunited to fish, share stories, make new friendships, and to trade. They speared salmon from the rock ledges and trapped salmon in weirs of woven brush. Salmon, shad, alewives, striped bass, and sturgeon all concentrated at the falls where the channel narrowed and the islands began.

Just below the surface at the hallowed fishing grounds, red-ochre cemeteries linked the eighteenth-century Penobscots to their distant ancestors; in their view, the fish themselves represented ancestry:

> Before there was a river there were streams, from the upland into the valley. But one day, the water in the valley became a trickle and it disappeared and the people grew thirsty. A young hunter went to find out what happened. He entered the forest and walked for days until he came to the place where the streams converged, and there he saw a giant frog. The frog grew bigger and bigger as it lapped up the little streams. The people sent for Gluskabe, our hero. Gluskabe followed the trail and when he came to the frog he called out, "There are others who are thirsty, too. You must learn to share." "I won't

stop," croaked the frog, "because I am the biggest and most powerful, I can do what I want." Gluskabe pulled up a giant white pine, and lifting high over his head he brought it down, striking the frog on the back. The frog burst into a thousand pieces. The water shot up into the air and landed in the deep furrow in the ground the tree had made, and the water began to flow. And that is how the Penobscot River came to be.

The released waters streamed down the mountainsides to the sea. Driven by thirst, people rushed into the water and transformed into the different species of fish.

Where did you come from?

On the coast, the strangers were well positioned to catch most of the salmon in the bay. They recognized salmon, because they had been fishing their own populations of Atlantic salmon for thousands of years. Settlers knew salmon to be a royal fish that made for good eating. They also knew that salmon fetched high prices—at one time, North American salmon were viewed as a way to eliminate Great Britain's national debt.

Before 1800, the Europeans used hemp nets to snare salmon in the estuary, and the swift waters and shallow rapids above the head of tide. Then they adopted weirs, simple fixed enclosures of sticks and brush. Fish swam in at high tide and were trapped when the tide went out.

In 1811, Hawley Emerson of Phippsburg designed a more complicated, mazelike weir with three enclosures or "pounds." In 1815 he built one at the mouth of the Marsh River on Treat's flats that enclosed such a mass of fish its sides burst open. Weir construction could be elaborate, and varied according to the fisherman and the contours of the river bottom and shoreline, but all salmon weirs shared the same basic elements. A straight fence (the hedge or leader) extended dozens or hundreds of feet from shore into the river channel, typically where the current met an eddy. When the ice went out in April, the fisherman drove stakes of black spruce into the muddy river bottom, and wove it with alder or birch brush, or else strung tarred cotton netting or six-inch wire mesh between the stakes. At the end of the leader was a series of heart-, arrow-, or double-curve-shaped pounds; a final enclosure had a board floor and a net to hold the trapped fish. Migrating fish swam along the leader into the pounds. Similar structures, made only of netting, were called pounds or traps, and were suspended in the bay by floats so they moved up and down with the tide. Pound-net fishing was practiced along the east shore of Islesboro, and from Camden north to Lincolnville, along the north shore of Ducktrap Harbor, to Northport.

A single weir could catch hundreds of salmon. By the middle of the nineteenth century, the Penobscot bristled with some two hundred weirs. They clogged the river where it split around Verona Island at the head of the bay; nets draped the islands and peninsulas, their ragged shores crusted with fish scales.

After harvesting, fishermen salted the salmon in barrels or smoked it. Smoking involved splitting the fish, removing the backbone (but leaving the head on), and salting for two or three days before tying the fish to cedar or spruce branches and hanging it high in a smokehouse for another two or three days.

Settlers caught hundreds of thousands of fish below head of tide and shipped them south. Their nets trapped the fish before they reached the falls.

The strangers began to settle. They swapped islands like stories and built walls across the falls, because good fishing spots, the lake outlets and waterfalls, were also where the current was fast enough to spin wheels and sawblades. The strangers cleared the woods, they took fish out of the river. And when they kept taking fish, taking and taking and taking, the natives protested. And when the strangers dammed the river and blocked the fish migrations and flooded the fishing grounds, the Penobscots, hungry, begged them to stop, "to prevent an evil so great it would be the total ruin of their tribe."

But the Penobscots were few and the Europeans were becoming many.

And the salmon that were so gratefully received by the native people and the strangers alike, where did they come from?

All salmon and other members of the family Salmonidae (grayling, whitefish, cisco, lenok, huchen, trout, char), descended from a common ancient species of freshwater fish. The oldest evidence of this ancestry, fossilized in lake sediments in British Columbia and Washington State, is *Eosalmo driftwoodensis*, an "archaic trout" that lived fifty million years ago.

Such fossils are rare, because the glacial landscapes preferred by salmon tend to be unstable, making the evolutionary history of Atlantic salmon difficult to reconstruct. The lineage is also confusing because, despite being separated for millions of years, *Salmo salar* shares much of its DNA with other salmonids.

The science is unclear on whether the common ancestor of Atlantic and Pacific trout was freshwater, marine, or both (diadromous). Some scientists have pointed out that it doesn't really matter. Fish have gone from rivers to the sea and back again for various reasons over millions of years. At some point, some of *Eosalmo*'s descendants began spending part of their lives in the ocean. Did they go to take advantage of richer food sources blooming offshore? Were they trying to avoid predators or competitors, or did changes in their surroundings force them to leave? Whatever the reason, populations of the sea-going fish became separated twenty-eight million years ago, as the earth cooled and dried out. The Bering land bridge emerged, forming a wedge between the Atlantic and Pacific fish.

Restricted to the ocean that became the Atlantic, *Salmo* began roaming farther, riding ocean currents to distant feeding grounds, staying at sea through the winter, returning to the waters of the pre-European continent. They became *Salmo salar*, while others that stayed closer to land became *Salmo trutta*, brown trout.

At some point before the start of the most recent Pleistocene era of glaciation, *Salmo salar* formed an eastern Atlantic (European) and a western Atlantic (American) population. For hundreds of thousands of years, as glaciers grew and shrank, grew and shrank, the salmon followed. Migrants established new populations in the shadow of the melting ice; expanding ice repeatedly pushed northern populations to extinction. Surviving salmon took refuge in the mid-Atlantic coastal rivers near present-day Virginia, and where now-submerged lands such as Georges Bank were exposed near the edge of the ice. For 200,000 years, Atlantic salmon lived with the ice, changing and adapting over time.

The last glacier melted away 14,000 years ago, and modern populations of salmon became established, moving one final time from multiple refugia into rivers that would become the Hudson, the Connecticut, the Merrimack, the Penobscot. These rivers are among the more than 2,500 draining into the Atlantic that once hosted salmon: from New York to Ungava Bay and Newfoundland, Greenland, and Iceland, and from Portugal to the Baltic Sea, Scandinavia, and Russia's Kola Peninsula.

Post-glacial pioneers repeatedly confronted with changing climate and sea levels, Atlantic salmon evolved a genetic program that is resilient to conditions many other fishes would find challenging. This program includes eggs that incubate through long, cold winters; an ability to grow fast during short summers; and flexibility in their life cycle (phenotypic plasticity) triggered by the often variable environments they experience: flashy water flows, wide temperature swings, and fluctuations in food supply. In this way,

Atlantic salmon are in a similar, if less overtly dramatic, situation as Pacific salmon, which have adapted to (and been periodically wiped out by) disturbances created by volcanic eruptions, landslides, fire, and flood.

Despite this resilience, adaptation to such extreme conditions comes at a cost. Salmon are, in many ways, more habitat specialists than generalists. They can't live just anywhere. Their complex, migratory life cycle is keyed to cold, productive oceans and the river landscapes created by glaciers. Many things have to fall into place to make such a life history work. And yet somehow they do fall into place, again and again, sending adult salmon back into the Penobscot River from the sea every spring.

Imagine you are thirsty. There has been no water for days, for years. Then there is water, and you are so thirsty you jump into the water. You drink. Your lungs turn to gills, your skin replaced with smooth, silver scales. You drink. You drink until you are full and then you emerge from the river. Which fish would you be?

Chapter Three

A Tradition Begins

Monday, April 1, 1912

Winter still held on much of the lower Penobscot River. Ice crowded the Bangor waterfront. Downstream from the narrows, blocked solid with ice, more than one hundred commercial salmon fishermen waited to drive their nets into the muddy bottom of the bay. The fishermen were anxious, unsure if the catch would follow the previous year's harvest of close to 9,000 salmon, the largest in years.

Early before dawn, Karl Anderson left his house and walked down Otis Street toward the river, where he kept his peapod-shaped fishing boat, a kind of double-ended rowing canoe. He made the boat himself out of wood and canvas, based on designs he remembered from his home country of Norway. Anderson had immigrated to Maine twenty years earlier, when he was nineteen years old. Since then, he had made his living and supported his wife and eight children as a house painter. Painting allowed him the flexibility to fish when he wanted.

Skunk cabbage poked through soft banks along Otis Stream. Pale sunlight, falling through the bare branches of elms and birches, coaxed thaw from stubborn patches of snow. Karl Anderson crossed State Street and walked past the hospital, across the railroad tracks, and beneath the new electric wires strung between poles along the tracks. Electricity had come to Bangor, and the dim days of gaslight were over.

Ice shelves clung to the shore along the high, turbid river. North of the hospital, a dam extended across the river from the brick waterworks that pumped drinking water to city residents. In the middle of the channel, the Bangor Boom held logs awaiting transport to the city's sawmills and shipping piers. The dam was built on rapids near the head of tide. Below the dam, ice had cleared from the salmon pool, where anglers vied each spring for who could catch the first fish, the biggest fish, the hardest-fighting salmon.

Competition was stoked by local hotel owners who wanted bragging rights to serve the season's first salmon to their guests. The Bangor newspapers reported on the first fish with details of weight, price, angler, buyer, and seller. Salmon were still big business on the Penobscot, selling for $1.25 per pound to markets in Boston and New York. Gallagher Brothers, Fickett's, and the Eureka Market all carried fresh salmon, shad, halibut, haddock, cod, flounder, and alewives. The first salmon of the year often graced the table alongside spring lamb at Easter dinner. Salmon sustained Benedict Arnold's Revolutionary War troops at Lake Champlain and, since at least 1776, when Abigail and John Adams served Atlantic salmon with egg sauce for dinner on the first Independence Day (because of its "American" quality), the region's traditional Fourth of July meal featured salmon with fresh peas and new potatoes.

"No fish of its magnitude brings so large a price per pound, and is so universally regarded as a chief delicacy," the U.S. Commissioner of Fisheries wrote. Maine salmon were "everywhere-popular," a valued and valuable culinary asset with metropolitan chefs and epicures. Salmon were often explicitly described as "Penobscot salmon," the premier and recognizable brand of American Atlantic salmon. Banquet menus listed "Penobscot Salmon, Sauce Hollandaise" for the fish course. Fannie Farmer's *Boston Cooking-School Cook Book* advised home cooks that "Penobscot River Salmon are the best."

Karl Anderson slid his cedar-planked, canvas-covered double-ender through the rushes between the ice-stained ledges into the cold water of the Penobscot, and navigated across the rapids to the Brewer shore and the salmon pool.

He threw a few practice casts. His fellow anglers watched from the porch of the Penobscot Salmon Club. Others gathered inside the clubhouse cabin, staying warm by the two fireplaces. They leaned back in woven chairs varnished to a bright vermillion, sucking cigars. Some may have been at the clubhouse since the previous night, ready to cook breakfast and be on the pool by first light. On the racks above their heads hung dozens of bamboo fly rods.

Since 1887, sportsmen from the surrounding valley and beyond had paid annual membership dues to belong to the first salmon club in the United States, and to be present on opening day of salmon season, "each with a setter dog, a pound of beefsteak and a quart of whiskey, the steak being for the dogs."

For years, anglers with means thought they had to go to Newfoundland or Labrador to fly-fish for Atlantic salmon, and for a long time there was even more confusion over whether or not salmon would take the fly at all, since adult fish do not eat after they enter freshwater. John Dymond conveyed the explanation of one experienced angler who suggested that on sighting a fly, the occasional salmon suffers an attack of temporary insanity. "This explanation is perhaps as satisfying as any other that has been proposed," wrote Dymond.

Fly-fishing was practiced on the Penobscot as early as the 1830s. In 1880, J. F. Leavitt and Hiram Leonard of Bangor were fishing in the Wassataquoik Stream, in the East Branch Penobscot River, when they caught a salmon on a fly. A few years later, Fred Ayer demonstrated that Atlantic salmon could be caught for sport right in his own backyard, where the fish congregated in the pool below the Bangor Dam.

Henry P. Wells, in *The American Salmon Fisherman*, reported that the Penobscot, as well as the St. Croix and Dennys Rivers, afforded some fly-fishing for salmon but was underexploited. Wells suggested that 1885 was the first year of real action at the Bangor pool, and credited the "quite respectable" catch to hatchery programs that had restocked "a practically exhausted river."

Fred Ayer caught six fish in the Bangor Salmon Pool in 1885; in the same season, Mrs. George W. Dillingham of New York became the first woman to take a salmon on the fly in the Penobscot when she landed a ten-pounder.

Faced with a shorter and less expensive journey than going to Canada, anglers from Boston and New York began traveling to Bangor in pursuit of the Penobscot's salmon. The biggest fish on record, at twenty-five pounds, was caught in the Penobscot in 1894. Angling for Atlantic salmon on the Penobscot River became the highlight of many a fisherman's dreams. Yet salmon were also within reach of the ordinary working man in central Maine. Men fished in suits during their lunch break; women set aside their hats and hiked up their skirts to fish. Children gathered on the rocks at the river's edge to watch—or else learned to fish from their parents, like Karl Anderson's children.

After a break for lunch and the turn of the tide, Anderson returned to the river. He cast his two-handed, eighteen-foot bamboo rod into the cold water, letting it swing downstream, moving the boat back and forth so the fly wavered in the current. The peapod shape of the boat allowed it to be maneuvered in the strong current below the dam. Anderson was competitive, but he knew the etiquette of scientific angling. He would not cast across a pool where someone else was fishing, or pull his boat between another boat and its quarry.

Karl Anderson caught two fish that chilly April afternoon. He was the only successful angler on opening day of salmon season in 1912. One of the salmon, a sixteen-pounder Karl fought for an hour, went to Campbell Clark, president of the Clark Thread Company in Newark, New Jersey, who frequently paid the highest price for the first fish. Karl decided the other salmon he caught that afternoon should go to the president of the United States.

Bangor delegates to the Republican State Convention had just voiced their unanimous support for the re-election of President William Howard Taft, and Karl thought that sending the eleven-pound, silvery-coated fish to Washington would "contribute to the city's need of honor and respect."

First fish rituals can be found wherever salmon live.

On the West Coast of North America, salmon are immortal beings. Every year, the salmon king orders them to clothe themselves in fish skins and go up the river, as a gift to the people on land and as a way to perpetuate their own kind. In honor of the gift, the people treat the annual arrival of the salmon in the spring with great reverence and ceremony. A fisherman selected, blessed, and purified by the tribal leader, catches the first salmon, which is then carefully prepared, cooked, and shared. The head is pointed upriver to show the salmon's spirit the way home. The bones are carefully cleaned and returned to the river, so the salmon can reconstitute itself and continue its journey.

The Ainu people of Hokkaido, Japan, regard the first salmon of the year as a sacred messenger of the divine. Servants of the Imperial household once hooked salmon from Tokyo's Sumida River and presented them to the Emperor as fish of rare delicacy.

The site of London's Westminster Abbey is rooted in salmon fishing: late one night in the sixth century, a salmon fisherman named Edric ferried a stranger across the marshy banks of the Thames River to Thorney Island, at the confluence with the Tyburn Stream. The man entered a new chapel that was being dedicated. While Edric waited in his boat in the darkness for the return of his passenger, he suddenly saw the windows of the new church light up and he heard the sound of chanting. When the man returned, he told Edric, "Tomorrow morning you must meet the King and the Bishop at the Abbey doors, bearing a salmon in your hands. Tell them that St. Peter has already consecrated the church on Thorney. Furthermore, in the future you must give a tithe of salmon to the Abbot of Westminster." When the king and the bishop entered the chapel the next morning, they knew from the crosses of consecration on the walls dripping with oil, and the remains of the candles, that St. Peter had been there. At this same site today, the Fishmongers' Company of London offers a salmon to Westminster each year.

Salmon fisheries were valuable properties in Great Britain; annual gifts of salmon were frequently among the benefactions bestowed upon monasteries by pious donors. In Gloucester, England, fishermen gave a thousand salmon each year to the lord of the manor. Knights celebrating a jousting victory called for a dish of salmon, which was cooked in wine, and brought by pages on a pewter platter to the sound of trumpets.

Spanish anglers auctioned the first salmon caught on a line during opening day to the best restaurants of Madrid.

In northern England, fishermen presented the first salmon taken from the Coquet River each season to the Duke of Northumberland.

Between ancient and modern times, traditions evolved, but the essential meaning of the first fish has not changed. Returning the first fish to the water, eating the fish in celebration, throwing back the remains, gifting the fish—all of the rituals show respect and honor to the sea as a source of life.

That the ocean pays back, sustains us, is suggested by the Latin name Carl Linnaeus gave Atlantic salmon: *Salmo salar*. *Salmo* derives from salīre, to leap, the name introduced by Pliny the Elder; *salar* means "of salt." Because salt once constituted a form of currency, *salar* also means salt money, an allowance, payment.

The presidential salmon custom maintained this legacy of seasonal celebration, while also bestowing honor upon a political leader, echoing the historic status of fresh salmon at the feasts of rulers and kings. Having rare and extravagant foods on the table was an expression of power, a declaration of the special qualities that set rulers and nobles—and their actions—above the common people.

The next morning, Karl took the fish to Oscar Fickett's market on Broad Street. Fickett packed the salmon carefully in a crate with straw and ice, and brought it around the corner to Union Station, placing the important package forward in the express car of the early afternoon train. On to Portland and then Boston, New York, and Philadelphia, the steam engine carrying the president's salmon whistled through the night.

The fish arrived at the White House the next day, Wednesday, April 3, 1912, and was received by housekeeper Elizabeth Jaffray and given to Alice Howard, the young cook in the White House kitchen, who began preparing that evening's meal.

Taft was the gourmand of American presidents. He ate a lot, especially when he was on the road and away from White House doctors who watched his diet. He ate venison for breakfast and salmon cutlets for lunch. The Tafts liked to entertain, and did so often. Never ones for formality and often running late, the Tafts always flustered the staff with last minute changes. "How on earth should this fish be prepared?" the cook may have wondered. In her Swedish homeland, she might have cured the salmon into gravlax, but there wasn't time for such a lengthy process. She looked at the clock. Mrs. Taft was insistent that dinner begin on time. In the large state dining room, staff set the table and lit the fireplace, throwing flickers of light on the walnut-paneled walls.

Nellie Taft must have hoped the salmon dinner would recall better days for Taft. Just two years earlier, they had sailed the coast of Maine in the *Mayflower*, the presidential yacht, and Taft visited the Bangor House hotel. At that time, Taft was still on good terms with his friend and mentor, Theodore Roosevelt. But things had changed.

President Taft was in the study, preparing an important message for Congress about making the machinery of the government more economic and efficient. America was experiencing an "efficiency craze." The mentality of mass production had spread from manufacturing and had every thumb on a stopwatch; even government was expected to be more efficient. Rivers, too, demanded efficient management, especially in the West, where disagreement about how to satisfy navigation and irrigation needs had contributed to the rift between Taft and Roosevelt.

In their designation of more than a hundred national forests covering nearly 175 million acres, Roosevelt, Chief Forester Gifford Pinchot, and Secretary of the Interior James Garfield had expanded executive authority over public lands and rivers. Fearing "the evils of monopoly"—large corporations like General Electric, Westinghouse, and Aluminum Company of America that had been taking over water, mineral, and forest resources—Roosevelt had been determined to maintain public control over the country's resources. Roosevelt had signed the General Dam Act of 1906, giving Congress (under Army Corps of Engineers advisement) the right to approve dams on navigable rivers in federally controlled lands. Roosevelt felt that the rule would put some order into a generally chaotic procedure, using his power to protect what he believed was public interest. Throughout the final year of his administration, Roosevelt vetoed every bill passed by Congress to authorize specific dams.

Private land and business owners throughout the country had been losing control over decisions about dam construction and water flows, and feared that centralized planning would obscure their own specific needs or projects. Water power interests had looked forward to Taft's presidency, and what they hoped would be more lenient regulations.

As Secretary of War under Roosevelt, Taft was in charge of recommending approval of new hydroelectric dams to Congress, and he opposed the restrictions enforced by Roosevelt and Pinchot. Taft, more prone to golfing than the fishing and hunting exploits of his predecessor, revised Roosevelt's policy by signing numerous bills granting perpetual and unlimited franchises for dam construction. He replaced Secretary of Interior James Garfield with one-time mayor of Seattle Richard Ballinger, a man who believed conservation was pointless without development, and that the federal government should give water power rights to private interests. When Pinchot criticized the new policies, Taft fired him, and then discontinued the conservation efforts, such as the National Conservation Commission, that Pinchot had initiated.

On Maine's Penobscot River, corporations such as the Penobscot Log-Driving Company held much of the power, and thirsted for more. As long as the mills and other riverside industries had been in operation, they had used river water flowing by their property. Extending a dam from a mill made it possible to power larger, more efficient saws and other machinery.

Nearly every stream along the Penobscot large enough to power a sawmill was dammed from one to a dozen points in its course. Walls of rock and timber were extended from shore into the river in order to funnel the current to spinning wheels. At first, with population sparse and mills few and far between, a mill owner could flood upstream land without much objection. But as population and land values increased, so did protests by upstream residents, leading to the first Mill Act in 1821 (based on a 1714 Massachusetts law) that limited water levels and impacts on other mills.

Mills had been considered a public good and a community benefit since medieval times, when they were often erected by feudal lords for public use. In essence, the only thing mill owners couldn't do was impact another mill's operations. Mill and dam development were subsidized by the work of government surveyors, who dispersed across the watershed to find more water power. In 1867 John Poor, A.D. Lockwood, and Hannibal Hamlin envisioned easy connections between the Penobscot River and Moosehead Lake and a demonstration project of dams and reservoirs along the entire Penobscot, the full employment of water power maintaining a population "as dense as that found in the manufacturing districts of England."

By the turn of the twentieth century, rivers had been repurposed as generators of electricity. Technology had evolved to transfer power economically over relatively long lines. In 1890, Bangor Electric Light and Power Company's electric station at the 25-foot-tall, 800-foot-long Veazie Dam became one of the first of its type in America. In the beginning of this new "electrical era," the potential for electric power provided by New England rivers seemed tremendous in relation to the existing and anticipated energy needs, an opportunity that had timber interests and other businessmen speculating for new dams and modifying existing dams for power generation. Many early hydroelectric dams were developed by industries, such as the mills, for in-house use during daytime working hours. Some of these business owners realized that they could make money by selling power generated at night or off-hours, and small utilities were born. But more power was being generated than consumed, prompting investors to promote their fledgling industry and create demand for their product. In Bangor, the electric street railway became the logical solution, in addition to residential and street lighting.

In the early 1900s, virtually every section of quick water on the Penobscot was surveyed for hydropower potential: Vinal's Landing, Birch Island Rips, Montague, Grindstone, Jordan Mills, Great Works, Mattaceunk Rips. One plan even considered diverting the water from Pushaw Lake through Caribou Bog to the Penobscot.

"[The Penobscot River] offers grand opportunities to the capitalist, through the development of its water powers," wrote one University of Maine engineering student.

"In traveling along the lower part of the Penobscot River a person interested in water power cannot fail to notice the absence of any concentrated fall of large magnitude in any one place," wrote another.

To address competing demands on Maine's waterways, the U.S. Geological Survey hoped to do for rivers what Pinchot had done for forests. Their study of the Penobscot River concluded that the watershed's lakes, ponds, rapids, and falls were underutilized. The engineers envisioned a "fully developed" river, each small dam doing its part to regulate flows and increase water storage in the lakes. The areas flooded as a result were of little value and were worth more as energy. In their report, the Geological Survey did not mention salmon, fish, or any other thing living in the river, and ignored the historic presence of the Indians, claiming instead that a large part of the Penobscot basin was heavily forested "wild land" known only to the lumberman and the sportsman.

Holding total faith that demand would follow supply, engineers surveyed and resurveyed every stretch of whitewater not already dammed. Where dams did exist, engineers drew plans for expansions. Just days after Karl Anderson caught the president's salmon, crews were preparing to start work on replacing the old wooden Veazie Dam with a 700-foot concrete structure at a cost of $150,000. "The old dam is not only out of date," the *Bangor Daily Commercial* editorialized, "but in addition leaks about 1,000 horsepower which will be saved with the new structure." The plans included two sluices for log rafts, but the newspaper made no mention of any kind of passage for fish.

A few miles below Veazie, the 625-horsepower Bangor Dam pumped drinking water to downtown residents, who could cook their salmon on an electric stove and eat it beneath the incandescent forty-watt glow of Mazda lamps, before heading out to the theater in the metropolizing downtown. The wires extended everywhere—the new form of energy, the abundant wealth.

By 1912, it was getting harder and harder for Karl Anderson and other anglers at the Bangor Salmon Pool to find a salmon, and the fish had gotten smaller, too.

Some blamed the mills and called for better fish ladders, demanding relief from the federal government's salmon hatchery programs. Federal fish commissioners acknowledged that opportunities for natural salmon production in the Penobscot were "exceedingly limited," because so many dams prevented the fish from reaching spawning grounds in the headwaters.

Some blamed the new pulp and paper processes, and the waste chemicals they dumped into the river.

Still others blamed the decline on commercial fishing in the estuary, which took thousands of salmon from the river compared to the few hundred caught by fly-fishermen each year. While many anglers ate their catch or made a profit selling their fish to Bangor markets, they saw themselves separate from the commercial harvesters.

The Penobscot Salmon Club anglers were among Americans who had found themselves in a rapidly changing world of overseas expansion, bulging factories, and technological advances. Doubt and unease persisted amid abundance and unlimited confidence. A feeling of over-civilization signaled broad dissatisfaction with modern culture in all its dimensions: scientific, technological, material. Rapidly evolving art forms reflected a fragmented identity: free verse in poetry, cubism in art; Marcel Duchamp's *Nude Descending a Staircase* captured the insurgency and bedlam felt by many.

The Arts and Crafts movement rejected the urban and industrial in favor of the rural and handmade, just as fly-fishing honored the craft of rod-making, fly-tying, and boat-building. Perhaps nostalgic for the days before the industrial revolution, men began to seek out "real" and intense experiences. The salmon angler represented the traditional, independent man many feared was disappearing beneath the tide of labor unions and corporate monopolies all heeding the "Gospel of Efficiency."

Fly-fishing was the opposite of efficient. Fly-fishing was a gentle art that stressed values other than material gain: honesty, fairness, hard work, and love of nature. George Bird Grinnell, editor of *Forest and Stream,* lamenting the passing of a more refined way of life under the assault of rapid industrialization, saw "correct" fishing as a way of differentiating the gentleman. Fly-fishing also required self-control in a time when so many aspects of ordinary life depended on decisions made by others in distant cities.

Fly-fishing had become a refined sport, with specialized equipment such as boats like Karl Anderson's peapod; hand-tied flies in intricate patterns of feather, fur, and thread; and the fourteen-foot, two-handed bamboo fly rod, perfected by Hiram Lewis Leonard on the banks of the Kenduskeag Stream in Bangor. Leonard developed a beveling machine to precision-cut uniform bamboo strips, eliminating the need for hand-planing and allowing production on a larger scale. Leonard's designs were refined by Fred Thomas, who eventually established the Thomas Rod Company in Bangor.

To lure and land an Atlantic salmon required practiced skill and participation in elaborate rituals. Beginners soon learned that their indifference became enjoyment, followed by zeal.

One authority wrote, "The lover of the 'gentle craft' who has never taken the salmon with an artificial fly cannot boast much of his professional skill, since angling for this magnificent fish is deemed the measure or standard of his capacity, the test of his art, the legitimate object of his loftiest aspiration. . . . There is no sport that will compare with scientific angling for exciting air, rambling over green meadows, in the grand old woods, among the rugged mountains, and over the beautiful lakes . . ."

Certainly, people who needed the food were angling for Atlantic salmon in the lower Penobscot. They caught as many salmon as they could, and ate them alone or with family. But these hungry folk were not writing articles for sporting periodicals, or joining the Penobscot Salmon Club, or lobbying for the passage of laws that would force all other fishermen to accept the "code of the sportsman": fish should not be killed in the breeding season or sold for profit; should be taken only in reasonable numbers, without waste.

Still, there were only so many fish to go around. Unsuccessful in stopping the commercial fishermen, and powerless against the industrial dam owners, anglers attempted to restrict anyone who was catching a disproportionate number of fish. Traditional ways of acquiring fish for food—such as subsistence fishing by the Penobscots—might have resulted in a greater harvest of fish at one time. Sport fishermen claimed that fish were being freely gaffed in secret spots inaccessible to anglers, that poaching was rampant.

The native Penobscots had developed many ways to catch fish: jigging hooks weighted with stones smeared with deer fat; tossing wishbone hooks on basswood line from hardwood rods; scooping with nets; harpooning; trapping with weirs and pots of woven ash splint; and casting hooks tied with tufts of deer hair or duck down at the end of alder sapling rods.

For salmon, *skamek*, big and silver and leaping upstream, they used spears. The Penobscots fished not the beginning of the salmon run in spring, when alewives, smelt, and shad were more plentiful, but waited for the peak of the migration in summer, when the fish were forty pounds fat.

From a ledge atop the falls and rapids, a waiting Penobscot fisherman would grasp the long spruce handle and plunge the *e'niga'hk*, or spear, into the water. Two dull hardwood tines on the outside grasped the fish so that the middle tine, made of sharpened hornbeam or iron, pierced its back. Sometimes the Penobscots fished at night from birch bark canoes. A bundle of folded birch bark tucked into the end of a long stick and lashed to the bow was ignited, attracting fish to the torchlight where they were easily speared, sometimes hundreds in a single fishing trip. The fresh fish supplied a feast; the rest were smoked on racks over the fire and stored for winter.

In Maine's capital, Augusta, a legislative committee had been charged with consolidating fish and game laws. As part of this exercise, and under pressure from the influential sporting clubs, in 1912 they reinstated an 1883 law prohibiting salmon fishing "any other way than by the ordinary mode of angling with single baited hooks and lines, artificial flies, artificial minnows, artificial insects, spoon hooks, and spinners," outlawing the use of spears to harvest salmon.

The "efficiency" of spearing may have prompted the Maine Legislature to outlaw the practice. Elsewhere in Maine, ice fishing had been restricted because it removed too many fish during a time when no tourists were present. Likewise, the legislature had already prohibited hunting in spring and summer. The new law levied a fine of $10-$30, plus $1 for each fish taken by "grapnel, spear, gaff with more than one prong." Spears were prohibited at camps, lodges, and fishermen's resorts, and could be seized if found.

Those Penobscots who wished to serve as hunting and fishing guides, something they had been doing for a century, also found themselves blocked by a law which required registration and certification by the state. All of these laws and rules, passed in the name of fish and wildlife protection, severed Maine's native people from their homeland and ways of life. The Maine Legislature followed national trends of increasing fish and game laws, and a desire to render Native Americans invisible.

And so the *American Angler's Book* argued, "It is claimed with some show of justice that the Indians have a hereditary right to the use of the flambeau and spear—it is the only way in which they take Salmon; but this is no reason why they should be permitted to practice it at improper seasons of the year, for the injury they do to the rivers is visited upon themselves as well as the whites, by the gradual extirpation of the fish."

U.S. Fish Commissioner Livingston Stone rallied in *Forest and Stream*, "There is not a cubic foot of water in the whole country where the salmon can rest in safety. The moment he comes in from the ocean he meets the gill nets and the pounds at the mouth of the river, the sweep seines further up, the hook everywhere, and at last on the breeding grounds, which at least ought to be sacred to him, he encounters the pitchforks of the white man and the spears of the Indian."

Many believed that indigenous cultures and beliefs were dying out according to the natural laws of progress and civilization, and that Indians should assimilate to, or would eventually blend in with, the rest of America. Someone who was going around at night with torches spearing fish would not blend in.

Twelve miles upstream from the Penobscot Salmon Club, the Penobscot Indians persisted. Poverty, cholera, influenza, and turberculosis had decimated their population to less than four hundred. With few fish to be had, some found that basket making and other traditional crafts provided better income than the farming encouraged by the government. Some Penobscots, because of their skill on the water, were hired to handle the batteaux and drive logs for the lumber companies. Many moved away.

The Penobscots remained surrounded by a landscape that was everywhere associated with their ancestors, their spiritual life, and their old stories. Should they begin to forget, the river was an ever-present reminder of who they were and who they had once been. Isolated in their main village on Indian Island, they hung up their spears and continued what traditional practices and beliefs they could.

In the Nation's capital, pink blossoms dripped from newly planted cherry trees, a gift from Tokyo that would flower each year, around the time that salmon fishing began seven hundred miles north in Maine. Tradition, however, was not forefront in the mind of President Taft. In April 1912, Bangor Republicans supported a vulnerable Taft in his campaign for the fall election. Roosevelt had abandoned Taft and the Republican Party to run again for president, in competition with his former ally. And oh, the stress! The stress caused Taft to put on weight.

The 332-pound president sat down at the great table in the state dining room before a platter of an eleven-pound salmon, the best size for eating. The fish, more delicate and less oily than salmon from anywhere else, was poached whole, bedecked with curly parsley, and served with a cream sauce.

Taft prepared himself to have a bite of something fine.

Chapter Four

Chesuncook

The Penobscot River watershed contains potato fields and cow pastures, towns and a few cities, mountains and hills, swamps, marshes, and hundreds of lakes and ponds. Most of it, however, is forest: dense stands of spruce and balsam fir; hills and valleys of birch, beech, maple, and ash; ridges of pine, hemlock, and cedar; bogs of tamarack, spruce, and beaver-dammed flowages of alder. The conifers give the land the look of the North, for the Penobscot Basin sits at the southeastern fringe of the great boreal forest that stretches from Newfoundland to Alaska, a post-glacial landscape governed by cold and ice.

Through this forest, which has been cut many times over, the West Branch Penobscot River flows east from a dam at the outlet of Seboomook Lake, turns north and then southeast, slowing as it approaches Chesuncook Lake. The river corridor is state-protected public land; the surrounding area, "unorganized" territory without local government, is the largest contiguous undeveloped space in the northeastern United States—but "undeveloped" does not mean uninhabited, or unaffected. Gravel logging roads criss-cross second- and third-growth forest; chainsaws whine from scattered campsites and lumber yards. All of this is hard to see from the water, which is edged on both sides by a tangled strip of alder, willow, and birch. Buffers of trees along rivers are known as "beauty strips," a term coined by loggers in the 1970s, when clear-cutting began on a large scale.

A canoe moves easily with the current, gliding over pools and skirting gravel bars. The West Branch, *Kettetegwewick*, looks like good salmon habitat, and it is, but today the salmon here are the landlocked variety, which were introduced to the river in the early 1900s (although landlocked salmon are native to Sebec Lake, which drains into the Piscataquis River, a major branch of the Penobscot). By then, dams associated with the logging and paper industries had blocked the migration of sea-run salmon into the upper reaches of the watershed. The alteration of the salmon's landscape began early; its legacy can be still be seen, if one knows how to look, in many places, such as Chesuncook Lake. Today it is hard to believe that salmon once traveled so far.

Every year, salmon migrate thousands of miles from the Labrador Sea and coast of Greenland, around Newfoundland and Cape Breton, along the coast of Nova Scotia and through the Gulf of Maine, into the Penobscot estuary, and inland for hundreds of miles. It is a feat worthy of contemplation.

Silver and strong from years spent feeding at sea, the fish enter the river in magnificent health. As spring turns into summer and the water warms, the fish move upstream in a drive to reproduce in the same streams where they hatched one, two, or even three years earlier. This ability to return to their home waters was noted in the seventeenth century by Isaak Walton, who recovered fish adorned with ribbons he had tied to their tails a year before, "which has inclined many to think, that every *Salmon* usually returns to the same River in which it was bred, as young pigeons taken out of the same dove-cote have also been observed to do." In 1873, Maine fish commissioner Charles Atkins attached aluminum tags with platinum wire to the dorsal fins of adult salmon after they had spawned, and released them. Some turned up above Bangor the following spring, showing that post-spawn adults sometimes stayed in the river over winter. Others returned two years later.

Exactly how they get home is unknown. Like many of the great migrants, Atlantic salmon have what humans might call an internal sense of direction. In the open sea, the earth's magnetic field, the sun and stars, maybe even the sound beneath the waves, can help them know where they are, and navigate home.

They travel south and west against ocean currents, thirty to sixty miles a day. When they reach the coast, they slow down in order to detect the right river and time their upstream migration to coincide with spring's increased river flows and warming temperatures. They travel near the surface, diving every so often, perhaps looking for currents carrying a familiar aquatic chemistry. They are guided, too, by smell, and recall information gathered when they first left their home waters. As juveniles, they "imprinted," or learned, the characteristic odor of their natal sites and waypoints along their seaward migration route. On returning, they draw on these memories, of rich minerals and the tang of pine needles, flooding from their olfactory nerve. Their desire like gravity, blind with lust, salmon charge against the flow.

Once in the river, they move fast through the estuary, reaching freshwater within hours or days. With extraordinary persistence and force, they make steady progress, swimming against swift rapids, climbing cataracts and launching themselves up unbroken falls as high as twelve feet. Within and around the fast waters, they find places to recover before the next push: deep pools and eddies, submerged snags of fallen trees, boulder shadows. The farther they have to travel, the more often they must rest, using lakes and pools between the rapids, cool dark places to pause on the long, tiring journey upstream. Only the strongest succeed, and sometimes only after repeated efforts. As they turn up the river's many tributaries, the scent differences become less and less, their senses more and more precise, searching for the one place they've known before, a certain tannin from a nearby bog, a fresh cut of eroded rock . . . once they find the spawning site, they hold for one or several months until spawning season in the fall.

Clearly, there must be an important reason why salmon desire to reproduce at the same location as their parents, a phenomenon called philopatry. One possibility is that if the habitat worked before, it must be a suitable place for making salmon—returning home is a kind of insurance that reproduction will succeed, the young will be adapted to the locale, and the family lineage will live on.

Salmon do wander, however. If they find the environment has changed, or the river is crowded with competing family members, salmon will colonize new habitat, usually close to their original home. But the pull of home is so strong, only two out of every hundred Penobscot River fish stray from their home stream.

Salmon migrating up the West Branch Penobscot River would have found that the river changed as they traveled. Approaching Katahdin, what was a wide channel with numerous large lakes became narrow and fast, strewn with large boulders and waterfalls. The leapers would wait in the pools on either side of the main drop, sometimes for months, gathering strength to propel themselves against the tremendous pressure of falling water. To leap up a falls, a pool below has to be deep enough to permit adequate acceleration—twenty-six feet per second—and the Leaper leaps, a light flashing through the mist.

For a long time, Europeans believed that salmon were able to get over such falls by bending their tails back toward or even into their mouths, and then releasing the circle, the force propelling them like a spring.

Nesowadnehunk Falls are just above the stream of the same name, which flows south into the West Branch after falling over Big Niagra; with a vertical height of twenty-two feet, Big Niagra Falls presented a natural barrier to migrating salmon, preventing them from reaching the waters that drain the west side of Katahdin.

After continuing up several more rapids on the West Branch, the river growing narrower, salmon would have reached Ripogenus, described by one traveler as "one of the wildest, most picturesque spots in the wilderness of Maine. . . . Rugged cliffs rising perpendicularly nearly one hundred feet on both sides, now with just room enough for the water to tumble through in fantastic, wild and beautiful shapes, again widening into immense gorges, into and over which the water rushes in myriad forms."

Another traveler wrote, "The water foams and hisses in its rapid course between walls of rock . . . in some places overhanging the stream . . . the river pitches through this gorge in a succession of rapids,—none very high, but together making a fall of two hundred feet or more."

Salmon made it through the gorge, where rock walls crowned with pine trees squeeze the river into a roiling chasm. At the top, they reached Ripogenus Lake, where they could rest. Another steep and narrow channel lay ahead before the flat, calm expanse of Chesuncook.

Historical reports suggest sea-run salmon came all the way up the West Branch, through Chesuncook and Seboomook lakes into the North and South branches of the Penobscot River, near the Quebec border. Millinocket, Nahmakanta, lower Nesowadnehunk, Caribou, and Caucomgomoc Streams were all good breeding grounds for salmon.

The same high-gradient reaches through which salmon traveled to reach the best spawning, nursery, and adult resting habitat also made logical sites for mills and factories that needed water power, a transformation that started with a desire for wood. Perhaps more than any other part of the Penobscot watershed, the West Branch was subjected to extensive dam-building, channel-straightening, and flooding in an effort to expedite the export of cut trees.

Prospectors first moved into the Maine woods in search of pine, which made the best masts for the king of England's ships. The 1691 Massachusetts Bay charter reserved all trees larger than twenty-four inches in diameter for the Royal Navy. Surveyors of pine and timber, of "His Majesty's Woods in America," marked the "King Pines" with a broad arrow, although not all colonists observed the mark, and most came to resent it.

Immense forests of giant white pine, *Pinus strobus*, over a hundred feet tall, six feet wide, and hundreds of years old, rose above the forest canopy and along the water's edge. The pines grew tall and unbranched for most of their length; resinous cones clustered in their crowns sparkled in the sun on clear, late-summer days. Their fine, feathery needles filtered light, softening the open ground below. Eagles and osprey nested in their towering branches.

But the strangers could only see the straight and true, a sea of spires. Pines became the symbol of the continent, their scent alerting weary sailors that land was near.

Pines were cut in the winter and teams of oxen hauled them to the nearest waterway. Around the time of the American Revolution, sawmills began crowding the river banks below Old Town, the "greatest scene of sawmill activity ever known in the state." Wealthy out-of-state businessmen like William Bingham purchased forest lands on speculation; mill owners sent "timber cruisers" into the woods to search for supplies.

Settlers moved inland, displacing the Penobscot people. They cleared lots and selectively cut the surrounding forest. As trees were cut along accessible waterfronts, lumber harvesters had to go deeper into the woods to find big trees and streams that could carry them out of distant woods. Making a stream or river "driveable" required extensive investment, including widening and reshaping brooks, and damming small streams and pond outlets. Temporary barriers held logs back ("splash dams") and ramps allowed logs to be rolled over obstructions ("roll dams") and walls shunted logs into preferred channels ("runarounds"). Abutments and booms kept logs in the main channel and out of the swamps and backwaters ("logans").

Depending on the weather, log delivery could be halted by the Penobscot's naturally low summer flows, an unacceptable delay to the emerging lumber barons whose fortunes depended upon steady volumes of timber. So, with privileges granted by the legislature, their men built bigger dams on headwater brooks and pond outlets to hold back water and ensure a constant flow.

Water flow was controlled to even out the quantity in the stream at any time, and held until it was needed to flush logs downstream through twenty-foot gaps ("sluiceways"). Logging took place in winter, when the ground was frozen and covered in snow, making transportation easier. Come spring, logs landed on the ice and floated free as it melted, sometimes with the help of explosives. Other logs stacked in rows along the stream banks were pushed into the water on the thaw.

Once the logs were in the water, they spread out over several miles, and immediately began to lodge against the shores, catch on rocks, ledges, and the streambed, and pile up in bank-to-bank snarls. Logging crews, perched on the rocks at rapids and falls, would push, pull, or roll stranded logs into the channel ("picking the rear"). They would heave and pry on jams for hours, until "all at once the apparently solid surface would begin to creak and settle. The men would zig-zag rapidly to shore. A crash and a spout of water marked where the first tier of logs was already toppling into the current. The front would melt like sugar. A vast, formidable movement would agitate the brown tangles as far as one could see. And then with another sudden and mighty crash that could be heard for miles, the whole river would burst into a torrent of motion." Logs left behind collected along the bends and ledges. Too much flooding pushed logs into the woods, which later had to be extracted. Horses were sometimes used along the shores, especially along the gravel bars and low banks.

Log-driving companies moved boulders and dug channels between lakes to redirect flows. They dynamited streams into chutes, flattened falls, widened narrows, blasted tight curves straight. That is how it started: here and there, river bends unbent, boulders loosened, gravity resisted.

Logging operations reached the East and West Branches of the Penobscot River around 1830, including the forest surrounding Chesuncook. A small village became established at the head of the lake, where logs were collected and sorted. Today, Chesuncook Village has a year-round population of ten, a cluster of buildings, an inn, and a small graveyard at the northwestern end of Chesuncook Lake.

Most visitors are recreational—canoeists, anglers, and hunters on their way north, to the Allagash Wilderness Waterway and the St. John River. But to follow the route of the Atlantic salmon means to follow the river's true course, twenty miles down the length of Chesuncook Lake. The third largest lake in Maine, Chesuncook presents a barren sea. Horsetail clouds fill the sky above an unbroken horizon of pale green and dark jagged spruce.

Then the wind comes on the heels of the rising sun, stirring the surface of the lake. As the sun lifts higher, the wind whips blue swells into whitecaps. Waves slam the gravel shore, rattling piles of driftwood.

The lake outlet was first dammed around 1834. The resulting flooding left stands of dead vegetation along streambanks and lakeshores, known as "dry-kill" or *dri-ki*. Sticks, pieces of stump, entire trees stand stark against the steady wind. Chesuncook is a transformed landscape; a wild and yet unnatural place.

Maps and other records provide evidence for what the Penobscot landscape was like before extensive settlement—and manipulation—by Europeans. Visitors came on behalf of British and French rulers and later governors of Massachusetts and Maine to survey boundaries, inventory resources, make maps. From these sources can be gleaned a picture, however faded, of the river's former self.

In the native Penobscot tongue, part of the Algonquin language family, *Chesuncook* means "place where many streams come together" or "meeting place." The *-cook* ending indicates slow water without difficulty or danger; a shallow, grass-filled deadwater, a kind of meadow in the river. Chesuncook was an important intersection where canoe travelers could head north to the St. John River, west to the Chaudiere, or south to the Kennebec. Along the sandy shore, archaeologists found evidence of the largest interior ancient village north of Bangor.

Joseph Chadwick traveled the West Branch at the behest of Massachusetts Governor Francis Bernard in the summer of 1764. Since much of the watershed was the domain of the Wabanaki, strangers like Chadwick needed native guides. The Penobscots were not happy about all the exploring, but eight of them went along to keep an eye on the white strangers. Chadwick's guides, including Chief Joseph Aspegeunt, wanted to ensure that Chadwick would keep his promise not to draw any maps (he broke the promise, sketching from memory after his return). In a little notebook of eleven pages, Chadwick described Chesuncook Lake as shallow with a mud bottom, surrounded by large tracts of grassland, higher intervale lands, and "sundry small islands all of which with the shore are good tracts of lands for a settlement."

Moses Greenleaf, Maine's "self-appointed publicist," created the first detailed and accurate map of the district in 1816. Greenleaf called Chesuncook "a fine sheet of water."

In 1820, the year that Maine became a state and separated from Massachusetts, first Governor William King hired an adventurous Major Joseph Treat to explore the outer reaches of the newly delineated state, to take inventory of what was still Wabanaki homeland, which had become a disputed territory between Maine and New Brunswick. Treat drew the river and its shores in fine-lined ink and cursive annotations, marking rapids and portages, ridgelines, rich soils, stands of pine and hardwood.

Treat was no stranger to the Penobscots. A major enterpreneur in the Penobscot Valley, he had leased lands from the Penobscots for cutting timber and hay, and—in violation of the 1796 treaty with Massachusetts—purchased all of the islands below the falls at Old Town, including Shad and Pine islands, the two best fishing locations, and Treat-Webster Island (known as French Island today). Treat competed with the Penobscots for fish, and built a small wooden smokehouse on Shad Island, a structure that some Penobscots later tore down in protest of what they saw as illegal occupation of their territory.

Keeping their enemy close, Penobscot Indians guided Treat up the West Branch in October. The water was low. They had to carry around much of the river, including a three-and-a-half-mile portage to avoid a succession of falls and strong rapids, rolling waves squeezed by walls of slate where the river gushed out of *No-lan-ga-mock* (Ripogenous) Lake, one of the most difficult portages on the Penobscot.

In times of high water, canoeists avoided the difficult carry around Ripogenus Gorge by going three miles up Nesowadnehunk Stream, through Kidney, Daicey, Beaver, Grassy, and Slaughter ponds into Harrington Lake, then downstream to Ripogenus Lake, or else via Mud Pond or Cuxabexis into Chesuncook. The low water conditions made this route impossible for the Treat expedition.

At the upstream end of Ripogenus Lake, a small and "handsome" pond, they carried their boats half a mile through another rocky gorge and around a steep waterfall at the outlet of Chesuncook.

Blowing winds forced the party to cross Chesuncook and reef their sails, "the wind then blowing a gale from southwest we could not cross the head of the pond to the inlet or river—arrived here at 1 p.m. and remain the afternoon the wind blowing hard." Despite the wind, Treat described Chesuncook as a "beautiful pond" with gravelly shores lined with "fine pine, rock maple, and yellow birch."

They stopped near a well-known native campsite. Treat noted that the shore and bank were almost covered with the bones of moose, deer, caribou, and beaver, signs that the area was an important hunting ground for the Penobscot tribe.

Treat had intended to go up the West Branch to its source, but the water was too low, and instead the party traveled from Chesuncook to the St. John River and back to the Penobscot via the Mattawamkeag. At a side trip up the Aroostook River, Treat purchased two salmon from an Indian for fifty cents.

Within two decades of Treat's expedition, the waterfall at the outlet of Chesuncook became the site of the first major dam on the West Branch. Future dams would eliminate much of the low flows experienced by nineteenth-century travelers. Creating impoundments changed the soundscape of the river valley, replacing the low rumble beneath the roar of rapids with silence.

To access more of the river's flow, mill owners had to build longer and taller dams. Production became more and more mechanized, requiring more water power concentrated in a single location—the kinds of places where fish also concentrated.

The first dam to stretch all the way across the river was at Old Town in the 1820s, flooding islands at the first big falls where the people fished. Great Works followed in 1830, Basin Mills in 1833, and Veazie in 1834-35. The 1830s saw a massive expansion of lumbering and related mill construction.

So many logging companies operated on the Penobscot, all anxious to get their logs down to the mills while the river was high from spring runoff, they got in each other's way, with violence resulting. In 1847, the legislature passed an act incorporating the Penobscot Log-Driving Company, granting "superior and controlling use of the waters of the West Branch," including the rights to clear the river of obstructions, to move boulders, maintain dams, booms, and other structures, to flood private property, "and to adopt modes and methods generally by which it could collect together great masses of water, and control and utilize them for driving all the logs in that branch of the river together." The company drove everyone's logs at once, sorting and delivering them to the appropriate mill. Some 30,000 men found labor each winter in the forests of Maine, including many Penobscots, who could apply their navigational skills and intimate knowledge of the river to running logs downstream.

Logs were driven to the Penobscot Boom above Old Town for sorting and delivery. Bangor was the shipping point for all the product sawed at mills in Orono, Veazie, Old Town, Stillwater, and Milford. The lumber trade hit its peak in 1872, when 250 million board feet from more than 400 mills passed through Bangor and thousands of wooden sailing ships carried logs, boards, shingles, spools, pegs, and toothpicks through the maze of fishing weirs in the estuary, and across the Atlantic to England, Australia, China, and the West Indies.

Tourists followed surveyors and scientists into the woods. When Henry David Thoreau visited Chesuncook in 1853, traveling in a nineteen-foot birch bark canoe steered by a native Penobscot Indian guide, the harvesting of the Maine Woods was approaching its peak. All of the forests within fifty miles of Bangor had been cut. Disappointed, Thoreau couldn't find the white pines he searched for because almost all of them had been removed. "Strange that so few ever come to the woods to see how the pine lives and grows and spires, to see its perfect success; but most are content to behold it in the shape of many broad boards brought to market, and deem *that* its true success!"

The dams on the river were not yet tall enough to completely flood the river, and the West Branch retained some of its natural seasonality. Travel could be difficult during a dry summer and fall, like that experienced by the Treat expedition. Similar conditions were encountered by a scientific survey crew that included fur-buyer, hunter, and naturalist Manly Hardy in 1859: "In many places, after picking out the large rocks with our hands and shoveling gravel with our paddles to make a channel, it took all three of us to drag a canoe to the next place where it would float. . . . At Pine Stream Falls, which is usually a smart pitch to run, and where batteaux have been swamped and men drowned, there was so little water that we had to drop the canoes down with a line. I learn that lately the raising of the dam at the foot of Chesuncook has backed up the water, so as to flow these falls out." Two years later, Hardy noted stands of dead timber that had been killed by flowage.

Nor were the dams tall or solid enough to completely block salmon migration. With their strong swimming and leaping abilities, salmon could surmount crude fishways and get up through leaky dams—as long as they weren't clogged or perched too high for salmon to leap into—where other fish could not. Relatively quickly, salmon came to be the predominant migratory fish in the river. And despite the dams and logging activity that was occurring on the Penobscot, the river was still less industrialized than other

New England rivers. The Connecticut, Merrimack, Saco, and Kennebec rivers had all lost their salmon runs by the early nineteenth century. America had lost more than half its Atlantic salmon habitat by 1850; by the turn of the twentieth century, Atlantic salmon would be eliminated from all but 4 percent of their historical habitat.

Salmon were seen frequently, and sometimes caught, at dams at North Twin Lake and Chesuncook. Salmon collected below North Twin Dam every spring, waiting for the log sluice gates to open. Salmon attempting to migrate hurled themselves upon the piers at Chesuncook Dam, and were taken by river drivers, with nets, gaffs, even axes. Others were injured or killed by the logs being pushed downstream.

But salmon continued to make it through. Manly Hardy reported taking salmon on a fly-hook more than thirty miles above Chesuncook in September 1873. River drivers noticed salmon in the West and East Branches, and trout-fishing tourists complained about salmon rising to their flies. Lucius Hubbard, author of the 1879 tourist guidebook *Summer Vacations at Moosehead Lake and Vicinity,* noted that good trout were often taken, and occasionally a salmon, in the pools below Chesuncook. In her journal of a canoe trip down the West Branch in August 1889, Fannie Hardy Eckstorm wrote about her father, Manly Hardy, catching a young salmon about eight inches long in the rips above Big Eddy, "a very handsome fish, bronze green on the back and sides, whitish beneath, with a row of dark thumbspots on either side and about two rows of bright vermillion spots."

The logging industry experienced a brief period of decline, until technological advances made it possible to make paper from wood pulp (as opposed to straw or cloth rags, which were in short supply). As pulp mills emerged along the riverbanks, at Great Works in 1882, and South Brewer and Orono in 1889, the North Woods harvest shifted from pine to spruce. Back into the woods went the crews to cut the forest anew. Forest cutting increased; the volume of cut pulpwood tripled between 1899 and 1930.

Hydroelectric technology emerged around the same time. "Improvements" intensified. Chesuncook Dam was rebuilt and more dams went up: Slide Dam on Nesowadnehunk, 1880; Millinocket Lake, 1883; Caucomgomoc Lake, 1884; Big Eddy, 1892; Seboomook, 1895. Below Chesuncook Dam, the river drivers blasted rocks to make room for Ripogenus Dam.

Five hundred feet across, the Ripogenus Dam had a log sluice forty feet wide and two hundred feet long. Sixty million feet of logs moved across the lakes and down the sluice. River drivers continued to vary the water level, closing the dam gates to float wood stranded after the winter, to hold the drive so logs could be unjammed, or save water for an upcoming drive. They opened the gates to move logs and water downstream in spring.

In 1900, the newly incorporated Great Northern Paper Company, with support from Standard Oil Company and the railroads, built a pulp and paper mill on the West Branch at Millinocket. The lumbermen, expressing the period's fear of monopolies and big business, resented the intrusion, but the corporation prevailed.

A 1,200-foot-long dam at the outlet of Quakish Lake diverted the West Branch through the mill to Millinocket Stream, turning practically all the water of the West Branch into power. The "Magic City in the Wilderness" produced 240 tons of wood pulp, 120 tons of sulphite pulp, and 240 tons of newsprint daily. Great Northern Paper rebuilt Chesuncook Dam in 1903-04, even as they were making plans to build a new, bigger dam at Ripogenus. They built dams at Dolby Pond, in 1907; Nollesemic Lake, 1911; Cuxabexis Stream, 1912; Loon Lake, 1914; Bear Pond, 1915; and Crawford Pond, 1916. Forty-eight dams went up on the West Branch of the Penobscot River between 1880 and 1925.

Great Northern Paper Company enlarged the Ripogenus Dam in 1916, merging Chesuncook, Ripogenus, and nearby Caribou Lakes and their connecting streams into one large, long reservoir. Great Northern took pride in the fact that their dams had "all but eliminated the extreme and lengthy flow variations commonly experienced under natural conditions." The water storage smoothed the hydrograph of the West Branch: peak flows were only three times higher than low flows. In contrast, on the Mattawamkeag, another branch of the Penobscot without storage dams, maximum flow was fifty times greater than minimum flows.

A dam forces a river to defy forces it cannot deny.

The water backs up—pushing, spreading, swelling, moving—it has to move. The river backed up above Chesuncook more than four miles, up Umbazooksus and Cuxabexis Streams at the head of the lake, and flooded out Pine Stream Falls. The raising of the water level, especially where the flow spread over wide, flat terrain, made the water too warm for young salmon. Rising water erased Native trails and campsites and flooded the meadows between river branches, killing trees and shrubs. The greater surface area of the lake increased the fetch, the distance over which wind travels without obstruction. The longer and straighter the fetch, the greater the wind and likelihood that big waves will form. The bigger waves, higher water, ice, and floating logs eroded the soft shores of Chesuncook, widening the lake even more, to about twenty miles long and two miles across. Dri-ki disfigured the soft grassy banks, flood-wood clogged the shores. Old logs and stumps weathered into white ghosts against the verdure. Today, wind rages across the open water, waves rattle the driftwood, bleached white by the sun.

Thoreau did not travel down the length of Chesuncook, but visited the village at the head of the lake before his party returned back the way they came. He spent the majority of his time observing the natives, botanizing, and looking for the elusive pine tree. More botanist than fisherman, Thoreau approached Chesuncook Lake "with as much expectation as if it had been a university—for it is not often that the stream of our life opens into such expansions."

Thoreau may have called this part of the North Woods a "bracing fountain of the muses," but today the fountain is dry. The flow has been altered, the natural river channel stolen, the route blocked by walls of concrete, the landscape made unfamiliar to the people, salmon, and others who had known it for generations. Salmon no longer migrate up the West Branch, but their descendants still carry a longing for home that haunts the displaced.

Chapter Five

Big Business

Thursday, April 6, 1916

Jeanette Sullivan of Bangor, fishing with her neighbor Patrick Nelligan, caught the first salmon one afternoon during the first week of salmon fishing season. She was the first (and only) female angler to catch a presidential salmon. Sullivan was well-known for her angling skill; her brother, Thomas Sullivan, oversaw fishing at the Bangor Salmon Pool as warden. She took her fish to Gallagher's Uptown Market, where throngs of people gathered to take a look at the president's dinner. A group of Bangor Democrats purchased Sullivan's ten-pound salmon, "the season's supreme delicacy," and Gallagher packed it for express shipment to the White House. It was President Woodrow Wilson's fourth salmon from the Penobscot River.

Wilson, a Democrat, won the 1912 election. He received only 42 percent of the popular vote but won an overwhelming majority in the electoral college, with Republican votes split between incumbent President Taft and "Bull Moose" Theodore Roosevelt. Intrigued by ideas and bored by details, Wilson proved an aloof but skilled administrator.

Mrs. Wilson brought her own cook from New Jersey to prepare family meals, but she also kept the Tafts's Swedish cook and housekeeper Elizabeth Jaffray. Broiled shad, a founding fish of the American colonies, was served at their first family dinner in the White House, but President Wilson, who grew up in Virginia and Georgia, preferred chicken salad, ham, peach cobbler, strawberry ice cream.

American attention had been focused on Europe. The assassination of Archduke Franz Ferdinand in Serbia in June 1914 released long-suppressed hostilities between nations, untangling treaties and dismantling alliances that had kept peace over the previous century. Within days, Austria-Hungary and Germany were at war with Serbia, Russia, France, and Belgium. Americans, shocked and uncertain, felt they had no business interfering on another continent and tried to stay neutral.

In Bangor, the "war scare" was blamed for a stagnant economy. "Where has gone the once prosperous river of Penobscot?" asked the *Bangor Daily News*. "Have our industries been paralyzed forever?" The editors lamented the decline of shipping and lumber and fishing; had they looked farther upriver, however, they would have seen that industry was alive and well on the upper Penobscot.

Per capita, Maine led the nation in developing hydroelectric power. More than households or commercial businesses, industry was rapidly consuming (and producing) the new electric energy. Since 1909, the large pulp and paper mills had added 3,728 kilowatt hours every year to fuel power plants. Through prior legislation and acquisition of other log-driving companies, Great Northern Paper Company had grown into the largest landowner and manufacturing corporation in the state, an "octopus" controlling much of the West Branch.

The concurrent rise of the pulp and paper industry and hydroelectric generation led to questions about who should control Maine's water resources. Almost everyone agreed that all available water power should be used; to do otherwise was wasteful and inefficient. Hostile toward foreign corporations, Maine wanted to keep its power at home, and generally opposed federal or even regional control. In 1909 the state passed the Fernald Law prohibiting the export of electricity generated within the state.

But should the government or private interests be in charge? The conservationist view was represented by Gifford Pinchot, who encouraged the Maine Legislature to keep the public's interest, telling them, "the measure of material civilization is the amount of power used per capita." The progressive ideology of Pinchot and Roosevelt continued to influence Maine's Percival Baxter, who served in the legislature and became governor in 1921. Baxter repeatedly tried to introduce legislation to give state government control of water power sites; he also tried to protect the lands around Katahdin. Private industry, led by the powerful Garrett Schenck of Great Northern Paper, defeated every one of Baxter's attempts.

To the industrialists like the paper companies, government control of hydroelectric power was a menace that destroyed individual incentive and responsibility. Great Northern Paper held up the West Branch of the Penobscot as an example of their bold willingness to develop water power all on their own. The company had just completed the Ripogenous Dam, an "immense concrete affair," a thousand feet long and eighty feet tall, impounding the "enormous quantity" of 22 billion cubic feet of water into the third largest storage basin in the U.S. and seventh largest in the world. "Creating this reservoir and storing a large amount of water which has heretofore gone to sea without being any benefit to anybody, will benefit every water power below the dam," declared the company. All they needed from the state was continued encouragement.

In 1917, the company introduced a bill that would amend the charter of the West Branch Driving & Reservoir Dam Company (which they had secured in 1903 after a long and costly fight) to allow it to sell or lease the right to Great Northern to take and use water stored by Ripogenus and North Twin dams for power development. In return, Great Northern would give up rights to transmit power to the public, which it held onto in its original charter.

To Baxter, Great Northern "took from the State of Maine what is perhaps the most valuable reservoir system in our state." He renewed his attempts to gain control of water resources on behalf of Maine citizens. He was halted by justices in Maine's courts, who determined it unconstitutional for the state to store, produce, or distribute power, especially because most of the power would be benefiting private enterprise.

Baxter kept trying. Lack of local demand and prohibition on transmission slowed hydroelectric development by public utilities, and minimal need for flood control prevented federal intervention, allowing Great Northern to proceed with development of the West Branch. In 1923 they got permission to raise Chesuncook Lake another four feet, over Baxter's veto. Governor Baxter's attempt to protect Mount Katahdin as a park was also defeated because legislators did not want to antagonize Great Northern, which had become so important to the economy of the state.

President Wilson's actions suggest that, for the most part, he would have approved of Great Northern's plans. He signed the bill approving construction of the Hetch Hetchy Dam in Yosemite National Park in 1913. He thought the fears of John Muir and others who opposed the dam were unfounded. Wilson approved a dam on the Tennessee River at Muscle Shoals to power the manufacturing of nitrate for ammunition and explosives. Yet Wilson also approved the creation of Sieur de Monts National Monument on Maine's Mount Desert Island, and signed the act creating the National Park Service, both in 1916, although in this he trusted his Secretary of Interior Franklin Lane to determine the best course of action.

War, and the scarcity and high cost of fuel and construction materials, had more people turning to water power, and raising the same questions people in Maine were asking. Should the federal government or private businesses develop water power? Wilson wanted resolution. He directed a committee to explore alternative procedures for water power decision making. Eventually, the Wilson committee's draft became the Federal Power Act, which created the Federal Power Commission and authorized the new agency to issue licenses for private hydroelectric projects on navigable waters or public lands for terms of up to fifty years. Congress enacted the legislation in large part to get away from the need to pass a separate piece of legislation each time a new hydroelectric dam was to be built, and, in part, to bring some coherence and efficiency to development of the nation's water resources. The act was disappointing and "feeble" to conservationists, who wanted greater coordination of river basin planning. The act had little impact on Maine, where most hydroelectric projects were above the "navigable" portion of rivers.

The United States entered the great conflict in April of 1917. Europe had been devastated in a manner that no one could have foreseen, and that American forces could hardly believe. Americans held nothing back, and Penobscot County citizens were ready to do their share. Fishermen signed on to the merchant marine and ship-building resumed in Bangor. Penobscot Valley farmers were urged to plant an extra acre of crops—potatoes, corn, beans, and turnips. Vacant lots turned into victory gardens. The University of Maine became a training camp, serving Penobscot River salmon with rice, tomato soup, peas, and fried potatoes to enlisted soldiers.

President Wilson appointed Herbert Hoover head of the Food Administration. A mining engineer and a skilled organizer, Hoover had helped 120,000 Americans stranded in Europe return home, and got food to the people of war-ravaged Belgium. In lieu of rationing, the Food Administration's first campaign asked citizens to cut back on meat, fat, sugar, and wheat and to participate in Meatless Tuesdays and Wheatless Wednesdays. They produced a book, *Foods That Will Win the War*, that touted fish as a protein substitute for meat. "Pound for pound, salmon, either fresh or canned, equals round steak in protein content." Salmon recipes included "Escalloped Salmon," a layered creamy casserole of canned salmon and crumbled crackers; salmon loaf, made with cooked salmon; and salmon chowder. Fresh salmon was among fish recommended for boiling, broiling, and baking.

The Food Administration encouraged fish consumption, but seafood had become more expensive as supplies grew scarce. Landings of shellfish and fish, including salmon, were down, but the fish business—fleets, processing plants, net and twine companies, fish oil factories—was thriving. Hatchery technology was more sophisticated. More people were fishing than ever before, with greater investment in gear, boats, nets, and weirs. Gasoline engines and steam power allowed them to fish longer and farther; with bottom trawls they could bring home more fish faster, and the higher prices led to better market conditions that "masked the mess."

The lights came back on with the war's end in 1919. It was a time of great emotion, but Wilson remained cool. His health had been failing since a stroke in 1919. He became withdrawn and preoccupied with international peace negotiations. The energy of the rest of the country was focused elsewhere. The public became disinterested in civic reform. Businessmen wanted to get the economy going again. They would find no greater ally than President Warren G. Harding.

Saturday, April 2, 1921

The river ran unusually high, and heavy rains clouded the water with silt. After a mild winter, more than six feet of water flowed ice-free over the Bangor Waterworks Dam. For the second year in a row, Michael Flanagan of 34 Pearl Street caught the first fish, a sixteen-pounder on April 2, one of five salmon caught that day. Also fishing successfully at the season's start were Maine fish commissioner E.W. Gould of Rockland, Albert Fischer of Fickett's Market, Charles Bissell, Karl Anderson, and Jeanette Sullivan.

Flanagan sold his fish to Gallagher's market for $2 a pound. "It is probable that Mr. Flanagan's salmon will grace the table of President Harding at the White House in the very near future, as a number of prominent Republicans were contemplating purchasing it and presenting it to the President with the compliments of the Bangor city committee," reported the newspaper.

Mayor Woods soon purchased the salmon for delivery, along with a letter announcing "the coming of the dainty" to President Harding.

Harding sent a thank-you letter in return, expressing appreciation for the thoughtful gift. He dined on the "delicious" fish at the White House on April 5, with distinguished guests Envoy Extraordinary of France M. Rene Viviani; French Ambassador Jusserand and his wife; Vice President and Mrs. Coolidge; and members of Congress.

After his victory in 1920, Harding and his wife, Florence, headed to Texas for some golf and tarpon fishing. On this trip he would meet with Gus Creager, a Texas oilman and land speculator, and Senator Frederick Hale of Maine.

Compared to Wilson, Harding was warm and old-fashioned, "an affable small-town man at ease with 'folks.'" He was the guy you wanted at the poker game. His food preferences reflected Main Street Ohio roots: knockwurst, pig feet, veal, pork loin. The White House food was mediocre and Harding social events lacked luster, despite ostentatious displays such as gold-plated silverware. Accessible to a fault, he longed for the good old days when the government didn't bother businessmen with unnecessary regulations.

The war had been over for two years and an economic depression was underway that would prove to be brief. A "Red Scare" was gradually ebbing, but the Ku Klux Klan was acquiring its first few hundred thousand members. The sins of flappers were about to disturb the nation; the first radio broadcasts crackled over empty airwaves. Prohibition had spread from Maine to the rest of the country; a few days after Flanagan caught Harding's salmon, law enforcement agents seized 75 cases of Canadian Club whiskey from beneath 375 sacks of potatoes in a rail car outside of Bangor city limits—after the car had made several stops to distribute wet cargo to thirsty customers.

Riding to office on a wave of reaction against wartime restrictions, Harding supported rapid resource development within an unfettered private enterprise system. Here was a man whom a country wearied of moral obligations could take to its heart. Spiritually, the nation was tired of all the talk about America's duty to humanity. They hoped for a chance to pursue their private affairs without government interference and to forget about public affairs. So they didn't flinch when the government sold oil drilling rights on public lands, and they didn't call for relentless investigation of the resulting financial mess. As Frederick Allen wrote in his classic snapshot of the decade,

"Disturbance of the status quo was the last thing that the dominant business class or the country at large wanted. . . . Americans wanted the President to let things alone, to give industry and trade a chance to garner fat profits."

During the war, fear of shortages and demand from Europe meant that natural resources began to flow as freely as ever into the hands of persistent exploiters. The government's priority was winning the war, which required tanks, ships, chemicals, and food. Demand for many products, including fish, increased during and after the war. Once scaled up for wartime needs, these sectors would not simply disappear, but would seek both continued government support and new civilian markets, and would therefore continue to look to the natural environment as a source of raw materials.

As a "raw material," Atlantic salmon had lost their place among Maine and America's important fisheries. Fannie Farmer had dropped "Penobscot salmon" from her cookbook. The average annual catch of Atlantic salmon in the Penobscot had fallen from 12,000 in 1890 to 1,500 in 1920. To nearly everyone outside the Penobscot River Valley and its angling community, "salmon" came from the Pacific, and usually in a can.

Atlantic salmon was first canned in North America in 1839 in St. John, New Brunswick, and in Eastport, Maine, in the 1840s, but the business was quickly overshadowed by the Pacific salmon canning trade, initiated in California during the Civil War. Maine fishermen built the first Pacific salmon cannery in Sacramento in 1864, based on their experience at home canning lobsters. The fishing industry soon spread north to the Columbia River. Shipments of fresh, whole Pacific salmon began arriving in Maine in 1884 when the first refrigerated railroad car direct from the West Coast rolled into Portland.

Within a decade, the Columbia River had become "a perfect web of nets," sending six million pounds of Pacific salmon to the East Coast each year, supplying the market after the last of the fresh Atlantic salmon had arrived from Canada. So much salmon was coming out of the West Coast that about half the annual production was exported, mostly to Europe. Ships carrying wheat and lumber brought canned salmon to Europe, establishing a market that encouraged more salmon canneries along the Columbia River, and more in Puget Sound, British Columbia, and Alaska. "Salmon are now rushed east from Oregon without being frozen and one was received at a local restaurant a few days ago which proved far more toothsome and delicate than the frozen western salmon which are usually served when the local fish are not in the market," noted the *Bangor Daily News*. The paper was quick to point out, however, that "Connoisseurs declare that the Penobscot salmon have a delicacy of taste and an absence of oiliness which is not obtained in fish from any other locality." By 1916, more than eighty plants were operating along the Alaskan coast, packing hundreds of millions of salmon every year. Cans of Alaskan pink and red salmon were readily available in Bangor markets.

The U.S. Bureau of Fisheries saw an opportunity to increase seafood consumption, and "made use of all available means to secure the fullest practicable utilization of the country's aquatic resources." In 1918, the Bureau obtained funding to build a fishery products laboratory in Washington, D.C., complete with an experimental kitchen and rooms for drying, smoking, canning, and refrigerating fish. Fish commissioners traveled the country giving lectures and demonstrations in preparing and cooking fish to housewives. By their own accounts, the outreach was "met with universal favor" but budgets ultimately ended the culinary tours.

Canned Pacific salmon had become "one of the staples of the world," and America's "most common fish food." New England fish processors responded to this competition by selling prepared salt cod, canned chowder, and salmon ready for salad or heating in cream sauce. Maine salmon remained primarily a fresh or cured product, however, because local canneries were busy packing lobster, sardines, clams, and white fish like hake.

Well accustomed to eating seafood out of a can, New Englanders readily adopted convenient, abundant canned Pacific salmon. Dozens of recipes for salmon loaf, salmon salads, and various salmon casseroles appeared in late-nineteenth- and early-twentieth-century community cookbooks.

President Harding's most salmon-friendly move was to appoint Herbert Hoover Secretary of Commerce. After the war, Hoover directed relief efforts in central and eastern Europe. He took over administration of the Bureau of Fisheries at a time when the Pacific salmon fishery "appeared to be hurtling toward collapse." Hoover, at the request of American canners, endorsed higher taxes on imported fish. He wanted the country's industries to fix their wasteful ways, and he believed government could help by coordinating planning. Hoover tried to get parties to find cooperative solutions, but he tended to side with industry in his focus on efficiency, standardization, and mass production. He supported science, but only when it demonstrated economic utility and benefited business, an attitude that drove fish commissioner Hugh M. Smith to resign in 1921. Fishers, canners, and pundits sent Hoover a steady stream of gifts, encouragement, and requests. Pacific canners inundated his office with fresh salmon and free advice.

Alaskan salmon fisheries had been steadily diminishing. When a conference on Alaska fisheries failed to find a compromise, Hoover began agitating for federal action. Hoover asked the president to designate fishery reservations over 40 percent of Alaska's coast.

At the same time, Hoover supported efforts to build major dams on western salmon rivers. He reconciled by restructuring the Bureau of Fisheries in favor of research on mitigative technologies: fish ladders and hatcheries became the order of the day.

On the Penobscot River, commercial fishermen had come to rely on the federally supported hatchery program, although their support ebbed and flowed according to the salmon migration. In bad years, they doubted whether the program was making a difference. But then a big run of salmon would arrive, as it did in 1912 and 1916, and they would renew their support. By 1919, local support for artificial propagation was again on the wane. Hatching salmon was expensive, between six and seven dollars for every adult fish. Questioning the wisdom of continuing the hatchery work, the Bureau stopped paying fishermen a bonus for live fish, resulting in many fishermen refusing to sell their catch to the hatchery and forcing the government to turn to Canada, trading trout eggs for Atlantic salmon eggs to augment collection of Penobscot fish.

The hatchery program was supported by Hoover, who had restored funding to the Bureau of Fisheries. Hoover had re-embraced Hugh Smith's approach to science, which supported a production-oriented view of rivers and fish. While conservation had gained momentum with creation of federal wildlife refuges and the Migratory Bird Treaty, fish were not considered wild animals in need of protection. Even Hoover's Alaska reserves allowed some fishing. Instead, aligned with the era's faith in mass production, fishery managers believed the solution was simply to make more fish. According to accepted scientific theory during the 1920s, the propagation function of the Bureau of Fisheries was both effective and necessary. Hatcheries and the bureaucracies that supported them became self-perpetuating, yet salmon continued to decline.

Hoover was a dedicated angler, and so he appreciated fish and supported government efforts to make more fish. "Fishing is not so much getting fish as it is a state of mind and a lure to the human soul into refreshment," he wrote. "But it is too long between bites; we must have more fish in proportion to the water." In his second inaugural address as president of the Isaak Walton League, Hoover laid out his theory that more nurseries were needed to raise fish beyond the fry stage, to ensure a better chance at survival when stocked. Fishing was a great equalizer, said Hoover, available to rich and poor. "We all have a divine right to unlimited fish." He promoted fishing as the best use of American's leisure time, rather than spending time in movie theaters, at baseball games, and other parts of the "machinery of joy."

The machinery, however, was becoming more and more dominant in American culture. Not a gear skipped when President Harding had a stroke and died two years into his term. Americans trusted that Vice President Calvin Coolidge would continue the policy of staying out of the way of business.

Tuesday, April 1, 1924

J. Edward Canning caught the first salmon of 1924 in the Bangor Salmon Pool on opening day of fishing season, below the new clubhouse that had been rebuilt after a fire in 1923. The sixteen-and-one-half-pound fish sold for $33 to Parker's Market in Post Office Square. The next day, ice filled the river, making angling hazardous. Jeanette Sullivan, who caught the first salmon in 1916, left Bangor that year. She gave her fishing tackle to her twelve-year-old niece, Roselle, telling her, "you ought to get lots of big salmon with this outfit." The next year, on her first fishing trip, Roselle caught a twenty-seven-pound "monster" salmon, the largest salmon taken from the Bangor salmon pool in twenty-five years.

With his New England frugality, Coolidge tried to bring back a sense of dignity and austerity to all aspects of the White House, including food. Coolidge inspected the White House iceboxes, criticized the menus, and once objected when the housekeeper prepared six hams for one dinner of sixty people. Coolidge was a nibbler—he snacked on nuts, crackers, and jam—and a practical eater. He made cheese sandwiches for himself in the pantry. At banquets, he was served roast beef in place of whatever fancy thing was on the official menu. An angler, he ate what he caught; the profusion of trout, pickerel, and pike that appeared in the kitchen after fishing trips, along with other fresh fish frequently sent as gifts, challenged the White House cooks to come up with sauces.

Coolidge believed conservation was sentimental. Factories were temples of worship and business the priority. Water power speculators renewed and strengthened their political influence. Electricity and petroleum fueled a second industrial revolution in America. Thousands of oil wells were drilled in Texas, Louisiana, Oklahoma, and California, oil to fuel huge generators, high-voltage transmission lines, and electric motors that brought cheap, efficient power to factories, office buildings, and homes. The spread of electricity cut production costs and created new markets for electric appliances as more and more American homes connected to the grid. Thanks to the assembly line, millions of Model T automobiles revolutionized American life.

Only wealthy and upper-middle-class Americans could afford electricity and the myriad items being cranked out by the machinery. More goods were being produced than the market could absorb, and so Americans needed to be persuaded to expand their desire for all the new products. Wartime propaganda for war bond purchases had proven successful, and the same techniques of mass-market advertising were applied to consumer goods. Large-scale manufacturing, through saturation advertising, could influence consumer choices (and sustain Maine's pulp and paper factories: Great Northern Paper produced hundreds of thousands of tons of newsprint, and was among the largest paper companies in the country). Brand names, equated with quality, convenience, and status, reduced the influence of independent businessmen like Bangor market men Fickett and Gallagher, who didn't carry the latest products. New products came forth each year that stigmatized last year's model, brand, or fashion as obsolete. They created markets where none had existed before.

Manufacturing continued to turn to science for assistance. Industry was reinventing food and putting new tastes on the market packaged in new ways, seemingly every day: Milky Way chocolate bars, Welch's grape jelly, Wheaties and Rice Krispies, Velveeta, bubble gum, sliced bread, Kentucky Fried Chicken, 7-Up. Scientific discovery of "vitamins" prompted encouragement to eat more varied menus. Fish was promoted as a healthy choice, wholesome and cheap. Housewives were supposed to know more about the food and medicinal values and fat content of different fish species, and were expected to try the latest preparations, such as frozen fish fillets.

The first U.S. patent for a fish-freezing process was granted in 1862 to Enoch Piper of Camden, Maine, who froze Penobscot Bay salmon by placing it on racks under pans of ice and salt. After about twenty-four hours, when the fish were frozen hard, he dipped them in water for a glassy sheen. Others improved on his process—fish was the object of experimentation because it was the most perishable. The technology spread west; by the end of the nineteenth century, British Columbia was shipping a million pounds of frozen salmon, halibut, and sturgeon to Europe annually. But most frozen food was poor quality—it was mushy, and tasted bad.

Clarence Birdseye revolutionized the frozen food industry in the 1920s, applying tools of mass production like conveyor belts to the flash-freezing of fish. His techniques were informed by his years spent trading furs in Labrador, where he discovered that the fish he and the local Inuit caught froze almost immediately after being pulled from the water—and still tasted fresh when thawed out months later. After starting out in New York, Birdseye moved to Gloucester, Massachusetts, and began freezing haddock fillets in rectangular cardboard boxes. In 1930 he introduced a line of frozen foods to retail markets, including fish fillets and Blue Point oysters.

Commercialization had a greater impact on New England's regional cuisine than declining resources. Stripped of the associations with unique places and cultures, fish and other foods carried generic labels that vaguely suggested the product's distant or exotic origin, contributing to the separation of people, place, and food. By sending a salmon to the president, the recreational fishermen kept alive the Atlantic salmon's status as both a product of a particular place and the king of fish.

On July 4, 1928, as tradition-minded residents of the Penobscot Valley sat down to meals of salmon and fresh peas, Calvin Coolidge went fishing on the Brule River in Wisconsin, with the aid of Chippewa guide John LaRock. He had announced that he would not be seeking re-election. To a relaxed and relieved Coolidge, life must have seemed pretty swell. The country's machinery was cranking along just fine, the stock market was ticking, and everyone seemed to be in a buying mood.

Hoover was so confident he would be the next president that he took time off from the 1928 campaign to fish for steelhead on the Rogue and Klamath rivers.

In spite of—or because of—his humble background, Hoover liked grandeur, and once in the White House he replaced the Coolidges's practical thrift with elegance and a "subtle yet rigid wall of social decorum." In part, this was because he hated social chitchat, and thought it a waste of time not to conduct business, even during meals. Lou Hoover helped shield her husband from personal engagement by increasing the formality of entertaining. She demanded perfection and efficiency.

Sometimes the White House staff were too efficient. When the directors of the Penobscot Salmon Club arrived at the White House to present Horace Chapman's fourteen-pound salmon in April 1929, and take what had become customary photos with the president, the fish had already been sent to the kitchen, where the cook had cut off its head and tail, and begun preparing it for the oven. The cook, an intelligent Irish woman named Katherine Bruckner, was equal to the emergency. She sewed the head and tail back on and stuffed the fish with cotton. A secretary brought the fish out to the White House lawn. "Hold it horizontal, it's fragile," he advised Hoover, but one of the photographers had already noticed a large piece of cotton sticking out of the fish. "Something is wrong," the photographer whispered to the president, who promptly covered the cotton with his hand as he held the fish up as dozens of photographers snapped away. "Just one more!" "Over here, just one more Mr. President!" The cotton kept oozing out of the fish.

Spectacle aside, the fish might have recalled Hoover's boyhood days, when he fished patiently in Iowa streams, and searched for trout in the mountain waters of Oregon. To Hoover, Chapman's fish represented repose from "the troubles of soul" imposed by the vast complex of industrialism. After a day of fishing, Hoover felt humbled, optimistic, and inspired, ready to return to civilization's cement pavements, office buildings, radios, telephones, church bells, piles of paper, and household chores.

Much of that civilization crumbled later that year, when the stock market crashed and sent the country free-falling into the Great Depression.

Hoover resisted direct federal intervention during the Depression, arguing that "it is not . . . the function of the government to relieve individuals of their responsibilities to their neighbors, or to relieve private institutions of their responsibilities to the public." While he continued some major federal construction projects, such as highways, waterways, Mississippi River flood control, and Boulder Canyon Dam in Colorado's Imperial Valley, he withheld his approval for the Coulee Dam on the Columbia River because of federal deficits and his nagging fear of federal power generation.

Major federal projects were unlikely in New England, where industries and population were on a long, slow decline that would never again catch up to other parts of the country. Lack of demand continued to prevent major developments. A 1927 amendment to the Rivers and Harbors Act directed the Army Corps of Engineers to evaluate navigation, flood control, power development, and irrigation on major rivers. On the Penobscot River, district engineer Colonel S. A. Cheney investigated nearly eighty potential power sites, including several of "considerable value" on the West Branch. He could not envision any expansion of mills or electric utilities. "The larger ones on the tributaries are in the northern wilderness where there are neither roads nor railroads and are so far distant from all markets that they can not be developed for many years." The Ripogenus Dam that Great Northern Paper boasted about just a decade earlier was "too remote to present economic worth," especially given competition from sites closer to industrial markets in states where exporting electricity was legal. While the dam-building days on the Penobscot had not yet ended, the peak of construction had passed, but not before taking a toll on the river's fish.

As federal fish managers struggled to force high dams to install fish ladders, Maine Commissioners of Inland Fish and Game were overseeing passage at fifty-five fishways in the Penobscot alone, and the last generation of commercial salmon fishermen was struggling to survive. In Maine, the legislature provided that the Commissioners of Inland Fisheries and Game could compel owners of dams and other obstructions above head of tide to provide durable and efficient fishways for salmon and other migratory species. The commissioners spent a good deal of their time inspecting fishways and following up on construction; in summer, they attended hearings about establishing the time and manner of taking fish, installing fishways and screens, regulating sawdust and mill waste.

Maine Commissioner of Sea and Shore Fisheries Horatio D. Crie of Castine vented his frustration in his report to the state:

> What are we doing to save the salmon from extinction except trying to enforce the laws? There are no adequate fishways and the ones we have are plugged almost continually with refuse. No small fish are being hatched from native Penobscot River salmon and practically all the salmon that get above the dam at Bangor go over the dam at extreme high tides. Such conditions are deplorable and must be remedied if we do not want the salmon to follow the sturgeon and shad which was once a valuable commercial branch of the fisheries.

Crie pleaded for more funding:

> The Penobscot River salmon are known all over New England as one of the best game fish, also one of the most palatable. . . . Are we going to let this valuable food supply slip from our last hold on it and go by default or are we going to give it the protection it deserves? The name 'Penobscot River Sal-

mon,' is cherished by every New England family and so let us have sufficient funds to protect it from extinction.

Beset by problems he could not solve nor even comprehend, Hoover retreated to his Rapidan River camp almost every weekend, where he spent hours building pools by hand in the stream, which he later stocked with rainbow trout. Hoover's thinking about natural resources was "curiously ambivalent." He favored conservation, except when it conflicted with his economy- and individual-focused philosophy. His primary contribution, achieved in spite of severe economic depression, was to rekindle national interest in the orderly development of natural resources. During his administration the era of executive laxity ended, the presidency once more became a constructive force, and the way was prepared for Franklin D. Roosevelt.

Chapter Six

Wassataquoik

The East Branch Penobscot River begins at the northern edge of the watershed, on the far slopes of Katahdin, and has always been the wildest country. The East Branch falls eleven times between Matagamon and Medway, a total drop of 408 feet in less than fifty miles: Stair Falls, Haskell Rock Pitch, Pond Pitch, Grand Pitch, Hulling Machine, Bowlin, Whetstone Falls, Crowfoot Falls, Grindstone, Meadowbrook Rips, and Ledge Falls. Salmon surmounted all of them in their desire to reach the spectacular spawning grounds, gray, cobbly islands and bars that also extended up the Wassataquoik and Seboeis streams.

But big business found its way here, too. By the 1830s large-scale logging operations were moving up the East Branch of the Penobscot. Logging began on the Wassataquoik in 1841, and tote roads pushed in toward Katahdin. The Telos Cut allowed logs to be transported down the East Branch.

In 1835 Mr. and Mrs. William Hunt established their homestead on the East Branch, across from the mouth of Wassataquoik Stream. They cleared land for growing crops, and fished for salmon in the river in summer and early fall. The Hunt Farm became a stopping place for woodcutters and river drivers, Indians and tourists, including artist Frederic Church and writer-tourist Henry David Thoreau.

Thoreau traveled down the East Branch in summer of 1857. According to his guide, Joe Polis, *Wassataquoik* meant "salmon river" and was applied to the entire East Branch Penobscot; the East Branch also translates as *Wahsehtek*, stream of light. At Hunt's Farm on August 1, they found the family had left but a few men were staying at the house while cutting the hay. "I do not remember that we saw the mountain at all from the river," wrote Thoreau, in reference to Katahdin. "I noticed a seine here stretched on the bank, which probably had been used to catch salmon."

Other travelers reported encountering salmon on the East Branch.

Louis Ketchum had dark, piercing eyes and high, strong cheekbones. A native Penobscot, he knew how to pilot a canoe, and he knew how to catch fish. In August 1861, Ketchum was working as a guide for investigators with the Maine State Scientific Survey, including naturalist Ezekiel Holmes, geologist Charles Hitchcock, assistant and Maine woods authority Manly Hardy, and young entomologist Alpheus Packard. They had traveled upstream from Old Town in a batteau and three canoes. The men were hungry and unhappy with their cook. Louis fashioned a spear along the way and caught what he could for supper. Packard wrote in a letter to his father, "We have had two messes of pickerel and eels, which Louis, our Indian guide has speared. We hope to get some salmon farther on . . ."

They went up the East Branch, and stopped for a few days at Hunt's Farm. The party split up, with one group going off to climb Katahdin and a second group staying at the river to collect mineral specimens. That left "Lewie" with some free time. In the evening he made a spear and caught two salmon, one ten pounds and the other seventeen pounds, which were cooked into a salmon and partridge stew. The next night, Hardy went along with Lewie. He wrote in his diary, "We went about two miles above the mouth of Wessaticook [Wassataquoik] and waited until after sundown. I shot a kingfisher and a muskrat; we then lit our torches and commenced to run down, I steering and Lewie standing erect looking like a fiend as the showers of burning bark burnt by the wood fell upon him." Lewie eventually speared a seven-pound salmon, which the cook boiled fresh and served with potatoes and coffee with milk. They moved farther up the East Branch and caught more trout, so many that young Packard was getting tired of them.

The East Branch Penobscot River is ideal spawning habitat for Atlantic salmon: wide, fast water, with lots of gravel along the river bottom, the kind of rubble that washes from melting glaciers and piles up in their wake. The river banks are sandy; today, the river has carved into the bluff at Hunt's Farm. Drift and eroded glacial deposits of sand and gravel are important sources of much of the Atlantic salmon's spawning and nursery habitat.

In order to reach spawning grounds and reproduce successfully, adult salmon need to enter the river fit and strong. Beginning sometime in the middle of October, when the water temperature dips below 45 degrees and the beeches have turned russet, those salmon that have made it to the breeding grounds take on elaborate changes in their physical appearance. The male turns mottled bronze and dark red, his lower jaw developing recurved teeth and lengthening into a curved hook (in contrast to Pacific salmon, who develop a "kype" on the upper jaw). Males arrive at the breeding grounds first, ready to fight over mates. They have not eaten in months in order to save their energy for reproduction. In response to hormonal changes, the stomach lining atrophies; liver, spleen, and kidney tissues degenerate; glands disintegrate.

The gravid female, now silver gray with bright crimson spots, has just a few days to deposit her eggs before they become overripe and infertile. She selects the nest site, usually at the downstream end of a pool, where the water deepens and accelerates into a gravel bar or riffle, sending oxygen welling up through the spaces between the stones.

She turns on her side and beats against the gravel to excavate a depression and clear out the finer particles. She may dig a few test pits before finding the right place. She releases hundreds of yellow-orange, pea-sized eggs into the loose gravel as the male, guarding, quivering, hovers next to her and releases sperm. Precocious (early-matured) parr hide in the nest, sneaking in to fertilize some of the eggs. Ten seconds and its over.

The female moves to the upstream end of the nest, or redd, and digs again, covering the eggs just deposited and creating a new depression for more eggs. She'll dig between four and fifteen pits over the course of a week and lay as many as 10,000 eggs. While not all adults die after spawning, most do, especially the males. In the 10 to 40 percent that survive, the glands and organs regenerate and grow again as if the animals had been reborn. The male's jaw shrinks; the red spots fade. The silver females still carry a few undeposited eggs.

Post-spawn adults are called kelts or black salmon. They move downstream in November, recovering by feeding on smelts and other small fish and insects. Some overwinter in lakes and deep parts of the lower river, and leave early in the spring. They can repeat the cycle a second or even third time, living an average of five years.

Burying the eggs in a gravel nest allows more offspring to survive, but it also means the salmon need clean gravel to fulfill their life cycle.

Salmon eggs incubate through the cold winter, and hatch in early spring. Spawning is carefully timed to ensure offspring emerge when conditions are optimal for the young. Salmon evolved this complicated life cycle because it worked: the juvenile fish grew up in protected streams with relatively few predators, surrounded by multiple generations of their family, but then left to spend their adult lives in the ocean where food was plentiful and nutritious. Overwintering at sea enhances growth, and helps to ensure that salmon enter the river in prime condition. Most do this early, in May and June, but such an early entry time results in lost feeding opportunities at sea. But in order for the cycle to work, the adult fish have to be able to reach their spawning grounds.

In long, steep rivers like the Penobscot, salmon have to enter early in order to have enough time to swim the hundreds of miles to the upper reaches of the river by the fall. They travel through the estuary and lower river after the high flows of spring runoff, but before the water gets too warm—by July, the water may have reached 70 degrees or more. But early entry also means the fish spend up to five months in the river before spawning, seeking refuge in deep aerated pools, springs, and tributary mouths and beneath boulders and fallen trees. Early migration leaves the fish vulnerable to predators.

Commercialization of Penobscot River salmon began well before the mass-production frenzy of the 1920s. The funnel-shaped estuary created a bottleneck, an ideal location to catch the most and tastiest fish as they made their way upriver, and the settlers fought among themselves over top tidal fishing locations. The estuary was the center of what became a major commercial salmon fishery, but salmon were caught for food throughout much of the main river and major tributaries by settlers and Penobscots. In the early 1800s, the Indians continued to camp in favored fishing and hunting locations in addition to their reservation lands, which in 1820 consisted of four townships on the upper river (near present-day Mattawamkeag and Millinocket), and the islands in the river. For the settlers along the rest of the river, fishing was a riparian privilege enjoyed by those who owned land fronting the water. Generation after generation fished for salmon and other sea-run species as part of the regular duties connected with small farms like Hunt's.

"Nearly every summer for many years there has been more or less fishing for salmon in the vicinity of Hunt Farm," reported Maine's fish commissioners in 1871. "Not a great many salmon are caught, for the fishing is only occasional, and with nets from sixty to eighty feet long, indifferently rigged. In fact there is little inducement to fish except for one's own supply, since purchasers are too far away."

When dams and other obstructions began to limit salmon migration, the people living upriver were the first to experience dwindling numbers of salmon.

In Europe, laws had been in place for millennia that limited catching salmon during their migration. Scotland restricted the fishing season in 1030. A Norwegian law, passed around 1200, stated that salmon should have free passage as far as they desired: "Running to mountains and shore God's gift should do, if running she wants." Massachusetts first enacted a fish passage protection law in 1709. In Maine, the earliest advocates for access to fish were the Penobscots, who believed the disproportionate system of fishing rights and sales of their fishing islands at Old Town violated their 1796 treaty. They and others, concerned about the no longer abundant fish, repeatedly pressed for regulation and the return of their sacred fishing places.

That salmon had to pass upriver to spawn was common knowledge, and so laws were passed, as early as 1786, requiring fishways at dams (for upstream *and* downstream passage). Towns could collect penalties from mill owners whose dams blocked fish migrations and restricted fishing.

In 1814 lawmakers limited nets in the Penobscot River to one-third of the width of the river channel. Citizens of Bangor, concerned about the salmon runs, appointed a committee in 1818 to address regulating the weirs.

The Penobscots regained three of their fishing islands at Old Town, but mill dams had already altered the water flows in the area. Penobscot Chief John Neptune asked the new Maine governor William King for help in 1820: "The white people take the fish in the river so they do not get up to us. They take them with weirs; they take them with drift net. They are all gone before they get to us. The Indians get none. If you can stop them so that we can get fish too, we shall be very glad."

Again in 1821, John Neptune and other Penobscot leaders went to Maine leaders "praying a law may be passed to prevent the destruction of fish in the Penobscot River." They pointed out that their "white brethren" downriver could use weirs and long nets, while they could only use "very small nets and spears."

In response, "expert" fishermen from the "big waters" in the estuary criticized the Indians for spearing salmon after they had "run the gauntlet and arrived unharmed at the still waters, where spawn is deposited." Even though all of the returning adults were destined for spawning, the weir fishermen thought that salmon only became "objects of solicitude" when actually on the spawning grounds. The lawmen should focus their efforts upriver, they claimed.

Several Penobscot fishermen were murdered for attempting to fish from disputed places; others were driven from the rocks, their nets destroyed.

Joseph Butterfield petitioned the legislature in 1822. He had lived in Old Town since 1803, when salmon, shad, and alewives were plentiful and the falls

> one of the greatest places for taking fish on the river, where the Penobscot Indians procured at least half of their living annually . . . now they cannot take a sufficient quantity for their families to eat even in the best part of the season and many of the white people who used to take plenty for their own use cannot get any by any means . . .

Butterfield attributed the disappearance of fish to the weirs in the estuary.

Testimony shows that the Penobscots did not want to stop all fishing, they did not desire *more* than anyone else, they just wanted their share. Hence the following story, in which Gluskabe learns a lesson in greed:

> Gluskabe went wandering about. When he returned to his wigwam, he saw his grandmother there fishing. He at last became impatient, as he saw that his grandmother was having a hard time fishing. Then he thought, "I had better help my grandmother, so that she will not have such a hard time." Then he made a weir across the mouth of the river, and left an opening half way in the middle, so that the fish could enter. Then he started out upon the ocean, and called everywhere to all the fish, saying to them, "The ocean is going to dry up, the world is coming to an end, and you will all die; but I have arranged it so that you will all live if you will listen to me. All who hear me, enter into my river, and you will live, because my river will survive! Enter all ye who hear me!" All kinds of fish came, until the fish-weir was full; and then he closed it up and held them there. Then he went to his wigwam, and said to his grandmother, "Now, grandma, you will not have to fish so hard, you will only have to go and gather as many fish as you want." Then Woodchuck went to examine what he had done; and when she arrived, she saw the weir brimful of all kinds of fish that were even crowding one another out. Then she went back, and said to her grandson, "Grandson, you have not done well by annhilating all the fish.

How will our descendants manage in the future, should you and I now have as many fish as we wish? Now go at once and turn them loose!" Accordingly he said, "You are right, grandmother, I will go and open up their weir"; and he went and turned them loose.

As each progressive fish passage law proved ineffective, numerous amendments were made changing the times, seasons, methods, and locations of fishing. Towns and counties established wardens to inspect and oversee fish passsage. But the attitude that upstream harvesters did more damage persisted.

Early laws forbid the catching and sale of fresh salmon after July 1, and prohibited all fishing for salmon year-round above the Piscataquis, although this was widely ignored. Natives and settlers upriver from the commercial fishery and the dams continued to harvest salmon, and to push for restoration of salmon runs.

Mill owners pushed back, believing destiny was in their favor. They gained exemptions from fish passage requirements on tributaries where they claimed fish were naturally scarce, the topography resistant to fishway construction, or, more important, the dams benefited the immediate community more than fish did. Allowing fish to pass would be ruinous, injurious, onerous, expensive. Mills needed the water; flow shouldn't be wasted on fishway "experiments." Often, the absence of fish was used as evidence in support of their argument against restoring passage. The mill operators believed their interest was more important to the regional economy. "Shall the enterprise and resources of the valley of the Penobscot be forgone for the sole purpose of saving the privilege of taking for a few days or weeks in the season a small supply of paltry fish?" To regulate mill operations was "so apparently obnoxious to the onward course of improvement on the Penobscot." Fishing was not an honest livelihood; fishermen were lazy and should be farming.

The fishermen saw the mill owners as wealthy, soulless corporations, and saw themselves as feeding the poor and hungry; but the bay fishermen clashed against the river fishermen.

By the 1850s, competition between weir fishermen, salmon netters, and upstream residents on the Penobscot had resulted in hundreds of confusing and contradictory special laws. But with so many changes in the regulations, sometimes multiple times in a single legislative session, and variable enforcement and compliance, it is hard to tell what worked or didn't work from year to year. From decade to decade, however, the trend became clear.

Catches in the Penobscot declined to a low point around 1860. In 1867, the Maine Legislature passed a "Resolve relative to the restoration of sea fish to the rivers and inland waters of Maine," restricting the locations, times, and days of the week when salmon could be taken from the Penobscot. The governor appointed Charles Grandison Atkins of Orland and Nathan W. Foster of East Machias as state fish commissioners. The legislature charged them with surveying the state's rivers to determine needs for fishways and fish propagation. In their first report, Atkins and Foster attributed the near extinction of fish in many streams to impassable dams, along with overfishing and water pollution.

Within a few years, Congress followed New England's lead and created a federal Fish Commission in 1871. President Ulysses S. Grant appointed Professor Spencer F. Baird, assistant secretary of the Smithsonian, as Commissioner. Baird noted that just a few years earlier, the suggestion of a possible exhaustion of fish populations would have been unimaginable. But the country was growing. Baird singled out Maine as one of the chief destroyers of fish, demanding study. "The interest in the preservation and increase of the salmon is due, in the first place, to its reputation as a game fish, and the sport experienced capture; but, perhaps, still more to its great size and economical value," he wrote.

Baird blamed impassable dams for the disappearance of salmon from so many rivers. "Excessive and ill-timed fishing" contributed to the decline and prevented recovery, Baird noted, "but fishing alone would hardly have sufficed to extinguish the brood of salmon in a single river. Commonly these two classes of destructive agencies co-operated. The dams held the fish in check while the fisherman caught them out."

Maine's fish commissioners were hesitant to restrict fishing because of the potential economic impact, "for fish that cannot be caught and used are of no value." They were also careful not to slander the mills, even as they noted smaller and smaller migrations. According to historian Richard Judd, the mill owners believed that manufacturing—civilization—was destined to replace farming, foraging, and fishing. They dismissed all competing claims to the river. That a brief, seasonal supply of "paltry fish" should hold up the enterprise and resources of the Penobscot Valley was inconceivable. Fishing was a waste of time, "a residue of backcountry indolence dragging down the valley's vigorous industrial economy."

Fishermen and their allies lashed back at the mills and the "wealthy influential few," invoking the region's fishing heritage. But the mill owners had an advantage: dams had already so diminished the fisheries that there was little left to protect. The U.S. Fish Commission concluded that "The interposition of artificial dams, unsurmountable by the fish, has been a great and perhaps the chief factor in diminishing the supply in [river spawning fish] . . .

The only way forward that state and federal commissioners could see was to make more fish by collecting spawn from wild fish and raising them in a hatchery. Baird agreed that "the art of artificial propagation" had a role to play in increasing the supply of food fish.

The state had been experimenting with relocating salmon eggs and fry since the 1850s, using culture methods based on European techniques. The world's first production-scale Atlantic salmon hatchery was built at Wilmot Creek on Lake Ontario in 1868, and the Sheepscot was the first Maine river to receive fish from this hatchery. Importing fish was costly, and the Fishery Commissioners of Maine, Massachusetts, and Connecticut asked Atkins to locate a suitable site on the river for raising Atlantic salmon. Their efforts eventually concentrated on the Penobscot River.

Atkins knew that moving adult salmon from one river to another wouldn't work, because the fish likely wouldn't stay put, try as they would to get home. But, since the salmon's homing instinct by then was well known, Atkins believed that if young fish were planted, they would return when grown to lay their eggs in the same streams. The easiest way to get young fish was to collect spawn from adult salmon and hatch them. One of the very first things the commissioners did, therefore, was "to cast about for a supply of spawn."

The East Branch picks up speed as it approaches the shallow end of a gravel bar, forcing water and oxygen down into the streambed, ten inches deep into the gravel, where tiny baby salmon—alevins—emerge in early May from a nest dug by the female the previous fall. The alevins absorb their attached yolk sacs before swimming out of the nest one night six weeks later. About one-third of the eggs survive to this stage. From the nest, the young salmon fry establish individual territories in shallow riffles with moderate current, but they stay close to their siblings.

They hide in the eddies behind rocks, in crevices, and amid submerged tree trunks and branches, facing upstream and watching for food to drift down, darting into the current for the larvae of mayflies, stoneflies, caddisflies; plucking black fly and beetle larvae from the bottom. They are already adapting to the specific temperature and flow regimes, predation, and availability of food in their home stream.

Fry grow slowly throughout summer. When they reach about two inches, they are called fingerlings or parr. Parr like it rough: faster water with coarse gravel, cobbles, and boulders on the bottom. Overhanging banks and trees provide a buffer against heat and predators such as mergansers and kingfishers. In the winter, the parr move closer to the shoreline, and are less mobile, staying in slower water between boulders or in the root wads of downed and overhanging trees, protected by stable ice cover. As they grow they move to deeper, faster water with larger gravel; they expand their territory—even using lakes—and eat larger and larger prey, including minnows and other small fish; and they defend their territory.

Out of every one hundred eggs, less than twelve will survive beyond one year. By producing more offspring than necessary, *Salmo salar* ensures that the young salmon best suited to their home waters survive to continue the next generation. In the ocean, the small fry and parr wouldn't stand a chance; but there are fewer predators in fresh water, so their chances of survival are much better.

Determined to find a supply of Penobscot salmon eggs, Maine fish commissioner Charles Atkins headed north in the fall of 1870, canoeing the West Branch from Mattawamkeag up to Ambejejus Falls, and the East Branch. He was looking for specific pools in the East Branch, ten to fifteen feet deep, just above the Hunt farmhouse and near the mouth of Wassataquoik Stream where salmon were rumored to lie. He did not find a single new redd. Driven off the river by wind, ice, and ten inches of snow, he gave up.

Atkins returned in November, traveling by road to Seboeis Stream. Bright patches of newly upturned gravel caught his eye. He looked closer, and saw the tell-tale excavation and the little mound of material below—a salmon redd. The next day he went down the East Branch, noting many new redds below the farm.

But with such a small number of salmon reaching the spawning grounds, Atkins realized he might have a better chance at success if he caught salmon in the estuary, where they were more plentiful, and somehow kept them alive until the fall. With cooperation from his fellow New England fish commissioners, he set up operations at the head of Penobscot Bay, at Craig Brook and Alamoosook Lake in Orland.

Several locations were tried for holding the fish until they were ready to spawn in the fall. Breeding was assisted by a "spawn-taker," who sat on a stool and squeezed eggs and milt out of the fish into a shallow metal pan.

Atkins produced 1,560,000 eggs in 1872, and stocked tens of thousands of salmon fry in the Penobscot. He was the first to pen the mature fish and keep them until their eggs were ripe, allowing cultivation on a grand scale. The hatchery came under the jurisdiction of the U.S. Fish Commission and was made permanent in 1889, the first federal fish hatchery in the country. The Penobscot River became the primary source of eggs for artificial Atlantic salmon production in the United States. The eggs incubated on plates of window glass in the basement of an old sawmill, and were shipped through the United States. Fish managers transported eggs around in moss-lined baskets. At one point a designated railroad car transported salmon around the state and country.

"Making salmon" quickly became an easy solution to a complicated problem. Chinook salmon, coho salmon, steelhead, and grayling were put in the Penobscot and other East Coast rivers (and in New Zealand and Germany). Penobscot fish went everywhere. Landlocked, or "Schoodic," salmon were spread to numerous lakes. German carp were placed in nearly every state and territory. Shad were introduced to the Mississippi and the West Coast. By 1900, Atkins had raised fifteen species at the Craig Brook hatchery.

Commercial fishermen supported the efforts. "They think the opportunities for natural reproduction are so limited that only a few years would elapse before the supply would become exhausted without the planting of artificially hatched fry," wrote U.S. Fish Commissioner Hugh Smith.

By 1879, the U.S. Fish Commission credited the work of Atkins with returning a profitable commercial salmon fishery, and providing food for locals:

> Salmon are no longer a luxury here, to be enjoyed by the rich only, but during plentiful seasons they are now often sold by our local dealers as low as ten cents per pound. Through the untiring efforts of the Fish Commission, for the past eight years, in restocking the Penobscot River, this once rare and delicate fish has been placed within the reach of the poor as well as the rich, and today the Penobscot is, in every respect, a salmon stream.

Recorded landings of salmon peaked in 1888 at 205,149 pounds. Improved methods of preservation and railroads made it possible to ship fish fresh from the sea to the country's interior. Demand increased in proportion. Salmon could be packed in boxes with snow, loaded in ice-filled refrigerator cars and delivered to New York in a day. Fresh fish were increasingly marketed, and to keep up each Penobscot fisherman had to build an ice-house and fill it with ice (cut from the river or nearby ponds) each winter. Most Atlantic salmon was shipped out of state. Wooden sailing vessels visited the Penobscot and other Maine rivers to load up with pickled shad and salmon. They bought salmon for six cents a pound, some already smoked and the rest they smoked themselves. The fish, along with lumber and miscellaneous produce, were delivered to ports in southern New England and New York.

Stocking, and efforts by fish commissioners to keep fish passageways clear, resulted in more salmon reaching the headwaters and more fish available for upstream residents. In 1879 more than one hundred were taken on the river at Hunt Farm and one was caught by canoe tourists at Matagamon; in 1880, 686 salmon were caught on the East Branch before the season closed on July 15.

The commercial fishermen profited from the hatchery work, yet residents on the East Branch and other parts of the upper watershed who took fish for sustenance continued to be vilified as vagabonds and criminals; an ignorant, drunken, vile crew of worthless poachers. Netting was banned above tide water "as a necessary measure to preserve the remnant of our fish."

In December 1885, E.M. Stilwell, Maine Commissioner of Fisheries and Game, wrote a letter to U.S. Fish Commissioner Baird:

> Above [Bangor] the waters are more shallow, the streams narrow and are accessible at all times day and night. The country is sparsely settled, and is infested by a class of poachers and outlaws, French, Indians and outcasts, who with net and spear and dynamite destroy every breeding fish they can reach. . . . It is the net, and spear, and dynamite only that can exterminate our fishes; fair angling never.

He also noted that "the freight on a single man coming into our State to angle for our trout or salmon or to hunt our game is worth a hundred times over the freight on the fish or game if sent to market and sold." Over time, the people above Bangor lost interest in preserving the salmon supply of the river. The salmon's shift from sustenance to sport was underway.

By the end of the nineteenth century, three of the five largest salmon populations in New England (Connecticut, Merrimack, and Androscoggin Rivers) had been eliminated. With dams continuing to multiply throughout the Penobscot River watershed, the number of returning salmon again declined, although the worst years still lay ahead.

After his success breeding salmon, and reports of large numbers of salmon reaching the East Branch headwaters, Atkins reconsidered an upper watershed egg collection station and again visited the East Branch. He estimated that about two hundred salmon had passed over the dams and spawned the previous summer, but the nests were scattered over some fifty miles of stream. To facilitate egg collection, Atkins proposed a weir that would direct fish into the cove known as Hunt Logan. When he returned to the East Branch in July 1896, he found one of the best spawning streams had been completely blocked by a dam, and the absence of salmon in the deep pool below the dam suggested that the run had ceased to ascend even that far. Again, in 1902, attempts to capture wild adult salmon by placing a weir across the East Branch, where salmon always spawned, were a complete failure because too few salmon surmounted the dams on the lower Penobscot due to the low water.

The Veazie Dam was especially concerning. Angler Karl Anderson noticed that a fish below the dam had a blunted nose, "as if it had vainly striven to get by some barrier." The Veazie Dam also encouraged poaching by "netters" or "drifters"—local residents were still trying to get their share of fish.

To Anderson, catching the fish was better than letting them drive themselves into the concrete, or turn red around the gills from pollution. Anderson called for better fish passage, and for stricter rules on the commercial fishermen in the estuary. "We have no objections to anyone making money from salmon fishing down river," he said, "but they ought to be subject to the regulations that are imposed on weirs in other rivers, chief of which is the provision that the weirs shall be open two days a week."

Salmon that were able to get past the dams still faced problems at the spawning grounds.

Rivers are in equilibrium with their surroundings. Forests regulate the flow of water, nutrients, and materials going into the Penobscot River, and the river reflects changes in the surrounding watershed. So when logging began around the West and East Branches of the Penobscot, the river was affected as much as the forests.

Unlike the Great Lakes region and the Pacific Northwest, where lumbermen adopted railroads as the most efficient means to transport logs to the mills, Maine lumbermen continued to rely on river transportation, and the manipulation of flows, well into the twentieth century.

Cutting trees along stream banks exposed the channel to hot sunlight in summer and frigid winds in winter. Soil dislodged by logging (and by clearing land for farms) smothered the streambed. Silt settled into the spaces between clean pieces of gravel, depleting oxygen, killing eggs, and preventing fry from hatching. Turbid water and rising temperatures drove out the mayflies, stoneflies, blackflies, caddisflies, and worms that salmon eat.

The dramatic alterations in flow associated with logging changed the volume and speed of water moving through the river channel and eroded stream banks. High flows flushed newly emerged fry and their food downstream; concentrated flows scoured pools where weary adult salmon rested. Discharge from sluiceways confused migrating adults. Sudden decreases in flow—especially in winter—exposed redds to freezing, stranded juveniles on exposed shorelines, and restricted the movement of parr and adults.

Reservoirs flooded out riffle habitat. With slower currents, food no longer drifted downstream. Where lands were not cut prior to flooding, the decay of the dying trees used up oxygen and increased acidity, killing salmon eggs and fry. In low pH water, older salmon can't regulate their internal chemistry; swelling in their layered gill tissues disrupts respiration. Acid also affects the abundance and variety of prey.

Bulldozing streambeds in preparation for log-driving, clearing vegetation along stream banks, and lining the banks with rows of stacked four-foot logs were common practices that disturbed the salmon's ecosystem. Moving logs scoured the stream bed, tearing apart redds and scattering eggs into the water.

Adult salmon making their way upstream and the juveniles heading downstream confronted not only dams but also heavy rafts of logs tumbling over the rips and clogging the sluiceways. Sawdust, bark, shavings, edgings, and other debris clouded the water. The rivers were full of wood. Logs covered the surfaces of lakes and booms (large rafts of logs collected for sorting and delivery to various mills), blocked light, and shed bark to the water and river bottom below. At one time, a three-acre jam of sawdust, edgings, and bark piled up at the Bangor boom. Log fragments floated out to sea; coarser pieces became saturated and sank. An estimated 4 percent of all logs driven down the Penobscot River—roughly equal to 400 million board feet of lumber—sank to the bottom of the river.

Much of the infrastructure associated with log driving was removed—though some was left for the forest to reclaim—but effects on the river persist to this day. Streams are wider and shallower than they might be otherwise. Log driving removed boulders and debris from the channels. Abandoned dams in remote sections of the Penobscot drainage continue to block fish passage, interrupt flow, and change temperature regimes.

William Converse Kendall spent decades studying salmon in the inland waters of Maine. In 1902 and 1903, he devoted his attention to the young salmon of the East Branch, including its two principal tributaries, Wassataquoik and Seboeis. Adult fish had been scarce—only a few adults made it above Grindstone—but Kendall found lots of young salmon, five to six inches and some ten inches.

He would not see them again. After state game warden Frank Perkins reported seeing salmon far up the East Branch in 1917, more than forty years would pass before the king of fish returned to the East Branch of the Penobscot River.

Chapter Seven

Empty Nets

Saturday, April 1, 1939

In the first silver flush of cold, damp Saturday dawn, scores of fishermen, bundled in woolens and hip boots and armed with long lengths of bamboo, sharp-nosed gaffs, and tins of feathered lures, moved down over the frozen shore of the Penobscot River to their peapod boats.

A heavy field of ice and snow choked the river, a turbulent gut of turquoise dwarfed by towering cliffs of gleaming ice. A dozen boats bobbed and pitched on the froth-streaked strip of open water as a blinding spring snowstorm tumbled out of the southeast. One by one, the anglers hauled their boats out onto the ice cliffs until only eight remained. It seemed useless, and as the stormy morning melted into a sun-splashed noon, all hope of catching a fish dimmed at low ebb.

Then with a rush, the first salmon of the season struck. Horace Bond was flicking his 5-0 double Atwood fly in Peavey Run, directly opposite the Penobscot Salmon Club, when the big silverside struck. Bond's light rod buckled and his reel screamed as the hooked fish darted for deep water. Other fishermen pulled their boats as far out of the way as possible to give the lucky angler plenty of room to fight his fish. The salmon broke water only once; a sparkling silver crescent momentarily framed in froth. After sixteen minutes, Horace Bond landed the thirteen-and-one-quarter-pound female.

Such is the story of the first salmon of 1939, as told by *Bangor Daily News* reporter Bill Geagan. The Young Business Associates of Bangor bought the fish, at $3.00 per pound, from Jones Seafood Market and shipped it by air to Franklin Delano Roosevelt in time for Easter dinner at his retreat in Warm Springs, Georgia. Polio-inflicted paralysis had forced Roosevelt to look at his environment in new ways, and Warm Springs was his place for recovery and healing.

Although the salmon was Roosevelt's seventh, reporter Geagan had faith that "as President Roosevelt is an ardent fisherman, the gift will perhaps appeal to him even more than his predecessors."

The timing was fortunate, for it meant the salmon would not be cooked by Henrietta Nesbitt, who had been brought to the White House from Hyde Park by the president's mother. Both women believed in the simplest of American cookery; they abandoned the traditional French cooking produced by the White House kitchen since it was first occupied in favor of "hearty, vitamin-filled dishes," such as roast beef and mashed potatoes. Nesbitt maintained that "a proper diet consisted of plain foods plainly prepared." FDR was something of a gourmet. He liked exotic fish and game, and gifts came in from all over the country: Lake Superior whitefish, Florida rock crabs, New England lobsters. Henrietta Nesbitt ruined them.

The food was so bad, Roosevelt built a new kitchen for himself on the third floor of the White House in protest. There, he could peacefully devour kippered herring and salt mackerel for breakfast, and chowder for lunch. Eleanor Roosevelt, equally fascinated by the gifts and with regional cuisine in general, equipped all the kitchens with electric stoves and dishwashers.

A few months after receiving the first fish of 1939, Roosevelt went fishing off the western coast of Newfoundland aboard the U.S.S. *Tuscaloosa*. He sailed again a few years later, to Ship Harbor, where he met Winston Churchill and signed the joint Atlantic Charter, which became part of the United Nations Charter. Afterwards, Roosevelt cruised to Penobscot Bay and stopped to fish, but with "indifferent luck." Souvenirs from these and other fishing exploits adorned the walls of the Congressional Waiting Room in the White House, where Roosevelt also hung a poster of big game fish and installed an aquarium.

Perhaps Roosevelt ate the salmon according to Nesbitt's recipe for boiled salmon (doubtful), or maybe he requested the king of fish be planked and roasted (his favorite). Given his preference for wild, fresh-caught fish, Roosevelt's personal cooks, Ida Allen and Elizabeth Moore, who had cooked for him in the Albany governor's mansion, likely knew just how to prepare the president's salmon. And as Roosevelt ate the fish, he would have savored every bite of a tradition that represented the best America had to offer.

With the Great Depression still raw, and another world war approaching, Americans in the 1930s favored austerity over personal or intellectual indulgence. Roosevelt felt that the Depression had brought a sharp break in American life. The New Deal was his way of saving what could be saved. Many of Roosevelt's New Deal programs attempted to reignite American pride as the nation rediscovered itself. The Civilian Conservation Corps sent young men into the woods, up to the mountaintops, and through the national parks to create spaces for this rediscovery. The Works Progress Administration (WPA), a sprawling operation that from 1935 to 1943 employed some eight million people in projects as diverse as sewing, drama, writing, road construction, and fly-casting lessons, had sentimental appeal.

These same feelings fueled the collective sense of urgency with which New Deal planners designed the landscape of a new age. Engineers, architects, and laborers sculpted progress into the Golden Gate Bridge, the Blue Ridge Parkway, Hoover Dam, Bonneville Dam, and the ambitious and unprecedented planning of the Tennessee Valley Authority, which included hydroelectric dams on nine major lakes. An undercurrent of environmental concern lay behind much of Roosevelt's administrative actions—at one point conservation measures accounted for nearly 11 percent of his federal budget. But what was considered "conservation" in the 1930s seems just the opposite in retrospect.

Massive infrastructure projects had the unintended consequence of destroying the very places and customs Roosevelt wanted Americans to be proud of—in the Tennessee Valley, thousands of families were forced to leave their land. The 726-foot Hoover Dam, the highest dam in the world, backed up the Colorado River for one hundred miles and generated hydroelectric power to enable the growth of Las Vegas and Los Angeles. New Deal water development policies set in motion an environmental transformation, interrupted only by war, which included the construction of more than 1,000 dams over the ensuing four decades.

The Federal Power Commission envisioned full development and integration of a New England power program through mini versions of the Tennessee Valley Authority dotted over five New England states, with Maine as the headquarters for more than half of the power supply. But most of Maine's landscape, largely under the control of pulp and paper companies, was spared.

Roosevelt likely believed in the wisdom of his engineers to ensure the survival of the king of fish in the Penobscot River. The WPA had built new fishways at the Bangor, Veazie, and Great Works dams in the winter of 1936-1937, and erected a new dam for a pulp and paper company at Mattawamkeag (the Weldon or Mattaseunk Dam) in 1939. A forty-foot fishway under construction on the new dam would be the tallest on the river. Raymond Dow, a member of the Penobscot Salmon Club, believed the fishway would clear the path for salmon from the sea to the East Branch spawning grounds. "Sawdust as a deterrent is a thing of the past. The commercial fisheries have been fairly well regulated by law," he reported. "If the present stock of fish can be maintained until the completed fishways permit natural reproduction, a much more effective condition will prevail than has in the recent past. A few years of heavy hatchery plantings at this critical period would go far toward putting the river on a self sustaining basis."

As it turned out, Atlantic salmon rarely used the Mattaseunk fishway.

Fishways are structures designed to help migrating fish move over and around dams. Fish jump up a series of stepped pools, or swim through slots in cross-walled chutes. In the case of the Denil fishway, named after the French scientist who invented it in 1909, fish navigated flumes with vanes on the sides to redirect water flow. The principle is that the fishway breaks up the vertical height, pressure, speed, and energy of falling water, and releases it at the downstream end in a way that mimics a natural current and attracts fish.

In the nineteenth century, with the exception of a few years (1833-1846) when the Veazie Dam completely blocked the river and the Bangor Dam was without a fishway (1874-1876), salmon could pass most dams on the lower Penobscot: they made it up through log sluices or side channels created by floods. They navigated fishways and surged over low-height dams with high tides and flows. The dams became less permeable as hydroelectric technology evolved, and fishways were both more necessary and more complicated, requiring the work of engineers and study by scientists.

The scope, scale, and speed of New Deal-era dam construction, particularly on West Coast rivers home to Pacific salmon, accelerated the evolution of fish passage design. And, by mandating engagement of fishery biologists with dam engineers, the 1934 Fish and Wildlife Coordination Act finally recognized that rivers could not be "developed" in a vacuum.

Construction of the Grand Coulee Dam on the Columbia River inspired the proliferation of trapping fish and trucking them around dams. At the Bonneville Dam, engineers and biologists created a $7 million series of fish ladders and elevators.

President Roosevelt had reluctantly approved a third fish lock at Bonneville, writing to Secretary of Interior Harold Ickes, "All I can hope is that the salmon will approve the spillways and find them really useful, even though they cost almost as much as the dam and the electric power development."

Interior Secretary Ickes, a lawyer, longtime advocate of better government, and passionate subscriber of Progressive ideals, including natural resource conservation, also had doubted the practicability of the fishways. So he was particularly interested in seeing them when he visited the dam.

"The run of salmon is over but, according to what we were told," he reported, "the ladders functioned perfectly during the run and the salmon had no difficulty in negotiating them to get to the immemorial spawning grounds above. Certainly they looked to me to be perfectly feasible."

As at the Mattaceunk Dam on the Penobscot, later studies revealed that even with the advanced fishways, the Bonneville Dam caused a few days delay for salmon moving upstream, and 15 percent of downstream migrating smolts did not survive passage over the dam.

Twentieth-century fishway designs attempted to work with the salmon's instincts to swim against the current. But engineers didn't know enough about these instincts, and one wrong factor—angle, location, distance between pools, water depth, number of openings, cross velocities or entrance eddies—could render the fishway useless. Fishway failures were the inevitable result of the attempt to mimic a system that evolved over thousands of years.

On the lower Penobscot River, WPA crews had to retrofit existing dams with new or additional fishways. At the Bangor Dam, the first barrier confronted by upstream migrating adults, salmon used the ramp only during times of high flow and low tide. Salmon had to swim seven feet into the fishway and make a four-foot jump at ninety degrees from a mass of rolling, turbulent water.

Those that succeeded then met the Veazie Dam, where a confusion of cross-currents inundated the fishway. One experiment found that of sixty salmon trapped and released above the Bangor Dam, only four swam beyond

Veazie. At the next dam at Great Works, flashboards (removable extensions to raise the height of the dam) blocked the flows in the fishway, which nearly dried up in summer.

Flashboards were also installed at the dam in Old Town to get more power when the water level dropped in summer. If the few salmon that survived the journey beyond the first three dams did not reach Old Town before flashboard installation, they were doomed to spend the summer in the polluted river below the dam.

Anglers in Maine, frustrated by the continued decline, took matters into their own hands. Members of the Penobscot Salmon Club and Penobscot County Fish and Game formed a committee to study the potential for restoration. In 1939, they hosted a conference with the Bangor Chamber of Commerce and U.S. Bureau of Fisheries that resulted in creating a Maine State General Salmon Committee, with Horace Bond representing the Penobscot watershed.

> It should be possible through careful study and adequate regulation to develop runs of a magnitude far beyond that visualized by most of the sports fishermen—runs which will be of great commercial importance to the New England states as well as providing extensive recreation for the sport fishermen. However, such runs cannot be developed purely through a hatchery and restocking program, for experience indicates that although hatcheries can be of great value in providing materials for restocking depleted streams and overfished areas, they provide no adequate substitute for natural propagation . . .

In 1941, a grand total of six adult salmon ascended the Bangor Dam. David Aylward of the National Wildlife Federation asked the U.S. Fish and Wildlife Service (USFWS; created when President Roosevelt combined the Bureau of Fisheries and the Biological Survey) to call a meeting of New England conservation commissioners, who soon realized they didn't know enough about salmon to draw up a logical management program.

The Penobscot Salmon Club helped state and federal officials capture more adult salmon for transfer to the hatchery. Nearly fifty million baby Atlantic salmon had been released in the Penobscot, but Charles Atkins and other biologists always said that while stocking could maintain salmon populations in a river with a relatively healthy run, stocking alone was not enough to re-establish a run that had been exterminated, or where stocked fish faced barriers of pollution and dams. When runs began to decline in the Penobscot during the 1920s and 1930s, Canadian populations (primarily from the Miramichi and Gaspé rivers) became the main sources of eggs for the Craig Brook Hatchery. But it was becoming more and more clear that something else had to be done.

With salmon still running, sparsely, in the Penobscot, on October 10, 1941, William Herrington and George Rounsefell of the USFWS; Maine's Commissioners of Inland Fisheries and Game and Sea and Shore Fisheries; and fishermen agreed to establish salmon propagation and stocking procedures, regulate the fishery, and coordinate research on Maine salmon as part of a long-range program designed to restore salmon runs to New England rivers. University of Maine students began studying salmon in the Penobscot River, and federal biologists made preliminary surveys of salmon habitat in a few streams and experimental plantings of Atlantic and silver salmon, but their work had hardly begun when it was curtailed by the Second World War.

Tuesday, April 1, 1947

White waves tossed the peapod boats of twenty hardy anglers. The sun rose, the chilly southwest wind eased, and anglers shed their heavy wool mackinaws and settled into rhythm: rod up, line back, rod forward, tip down, roll line to the current, brush the surface of the pool with the fly as the oarsman swings the boat side to side downstream.

The current quickened in the ebb tide. Tired fishermen beached their boats on the gray ledges and went inside the clubhouse for lunch. Two boats remained on the water. Below the Peavey Pool, Donald Smith began to reel in his line when a salmon struck, the first fish of 1947.

Local fishing guide Charley Miller offered to cook President Harry S. Truman's salmon meal with peas, potatoes, hot johnnycake, hot gingerbread with whipped cream, and "good coffee." Instead, U.S. Representative Frank Fellows, a Republican from Bucksport, presented the seven-and-three-quarter-pound fish to President Truman. During the ceremony, Fellows remarked that "getting this salmon every year is about the only reason he could see why anyone would want to be President."

Truman laughed. "There's something to that, Frank!"

Truman was neither picky nor greedy about food. He learned in the army to eat what was put before him and like it, although the Trumans eventually replaced Roosevelt-era fixture Henrietta Nesbitt in favor of someone who better knew their tastes for farmhouse cooking.

With many countries devastated and starved by World War II, Truman advocated for conserving resources as sound economic and democratic policy. In the first televised speech from the White House, he encouraged Americans to avoid meat on Tuesdays, eat no chicken or eggs on Wednesday, and save a slice of bread every day. As a result of wartime food shortages, per capita consumption of fish and seafood products had risen to its highest point in history, in part driven by the production of canned fish for troop rations. Canned fish became regular table fare in such dishes as canned salmon bake, tuna noodle casserole, and sardine sandwiches. At the White House, luncheon menus featured salmon croquettes and heated canned salmon, along with the occasional broiled salmon steaks. For the first time, canned tuna sales topped those of canned salmon, making tuna America's most popular fish.

The war changed the way Americans ate. Women returning to the kitchen after wartime work sought easier and quicker ways to prepare meals. When a shortage of metals led to a decline in canning, many people tried frozen food for the first time, and the frozen food—and supermarket—industry began.

Even those who wanted to eat a fresh Atlantic salmon would have had a hard time finding one. As President Truman was accepting the first fish of 1947, Penobscot River fishermen harvested a mere forty fish from their weirs, forcing the state to shut down the commercial fishery.

Rachel Carson, writing at the time for the USWFS, advised housewives to substitute fillets of wolf fish, an abundant and "underutilized" species, for "expensive fresh salmon." No salmon recipes appeared in the Maine Development Commission's recipe contest booklet published in 1945. Banquet chefs seeking whole fish turned to West Coast rivers and their plentiful runs of coho and sockeye salmon. The only way to eat a Penobscot River salmon was to catch it yourself or know someone who was an angler.

In some minds, the commercial fishery might as well have continued, since even those salmon that evaded the weirs couldn't reach their spawning grounds. The fishways so optimistically constructed by the WPA weren't working, and Maine's newly created Atlantic Sea-Run Salmon Commission knew it. Since the construction of the Mattaseunk Dam, very few Atlantic salmon had been seen in the fishway. A few used the ramp at first, but no one had seen a salmon in the fishway since 1945. In reality, Atlantic salmon could access only 2 percent of their habitat in the Penobscot watershed.

The Atlantic States Marine Fisheries Commission, formed in 1942 to coordinate the conservation and management of commercial fisheries along the Atlantic Coast, became actively interested in salmon restoration and urged Maine to end the overlapping jurisdiction of the Inland Fisheries and Sea and Shore Fisheries departments. In 1945, Governor Horace Hildreth appointed a special three-man commission to report on the situation, which they did in January 1947:

> We believe that the Atlantic salmon constitute a great natural resource now sadly depleted but still capable of restoration to the great financial benefit of our people. What is needed is a sound program to clean up pollution and river obstructions where those factors are detrimental.
>
> There is no doubt that commercial waste from industrial plants and sewerage from cities and towns along the [Penobscot] river have caused young salmon to die on their way to the sea. Then too, vast quantities of young salmon surely must be ground to death in passing through the whirling machinery of power plants. First we believe that the river should be freed of pollution. Then salmon ladders should be properly placed and measures taken to place protective screens of some sort in front of all power machinery.

The report stressed the need for a permanent salmon commission, which was created in 1947 and vested with the power to create rules for the recreational fishery, including special fishing licenses for some waters. The 1941 agreement between USFWS and the state agencies was redrawn and extended to include the Atlantic Sea-Run Salmon Commission and the University of Maine.

In their first report, salmon commissioners George Rounsefell and Lyndon Bond observed that eight Maine rivers had regular salmon runs, but the sport fishery was suffering "due to a great scarcity of salmon." Of the really large streams, only the Aroostook and the Penobscot offered any immediate hope for restoration. With the commercial fishery restricted, they hoped a vigorous campaign of stocking, fishway improvements, and pollution abatement would bring results in the Penobscot:

> The success or failure of any salmon restoration program lies in the condition of the streams . . . it is striking that with very few exceptions it has been the better salmon streams that have suffered most from civilization. The streams with the steepest gradient, the largest volume, and the most constant summer flow were best both for salmon and for industry . . . on the Penobscot River the salmon must pass over six dams—at Bangor, Veazie, Great Works, Old Town, Howland [West Enfield], and Mattaseunk—in order to gain access to the great spawning and nursery areas of the East Branch and the Seboeis and Wassataquoik tributaries. The present fishways, built after the dams were already

constructed, were necessarily a compromise and it is doubtful whether a permanent salmon run of any magnitude could be maintained until they are improved.

The commissioners believed in fish passage engineering. There was no limit to the height a salmon could be trucked or conveyed with fishways; if needed, an endless fish ladder could convey salmon from estuary to headwaters. The fishways just needed to have the right design and be built correctly, inspected, and maintained. They also believed in hatcheries, and restarted operations at Craig Brook.

After the Second World War, the concept of efficiency and scientific management, so pervasive when William Howard Taft received his salmon in 1912, had evolved into concepts such as "maximum sustainable yield." Managers stocked desirable fish and poisoned unwanted fish. They constructed and modified water bodies. They regulated fish harvest with the single aim of providing the greatest quantity of salmon so that the greatest quantity of salmon could be taken—by sport fishermen.

Not everyone felt this way. With the publication of *A Sand County Almanac* in 1949, Aldo Leopold called for a new measure of progress that would redefine "progressive" civilization as one that valued and perserved its remaining wilderness. Old growth trees instead of stumps, teeming fish runs instead of dams. Leopold called the fixes of the era superficial: "Flood control dams have no relation to the cause of floods. Check dams and terraces do not touch the cause of erosion. Refuges and hatcheries to maintain the supply of game and fish do not explain why the supply fails to maintain itself . . . the practices we call conservation are, to a large extent, local alleviations of biotic pain."

Truman negotiated fishing treaties with other North Atlantic nations, yet domestically he supported dams and water supply planning that viewed rivers as blank slates to be developed, floods in need of control, conduits to dilute and carry away pollution. Food was no longer a service Americans needed from rivers. With diesel engines, bottom trawls, sonar detection, and onboard refrigeration and freezing technology, fishing had moved out to the high seas. Motorized vessels rigged with chain sweeps and gillnets made more fishing grounds accessible.

With commercial "fishing" and "fisheries" moving offshore into the Gulf of Maine, the backyard Atlantic salmon runs were worth more in recreation-related expenditures than as food. The purpose of management was not to restore a commercial food fishery, or rejuvenate a river ecosystem, or return what had been taken from the native inhabitants. Those goals would come later.

Monday, May 31, 1954

It took two months for someone to land the first salmon of 1954. Flannel-clad Bangor resident Guy Carroll took the ten-pounder from the Bangor Salmon Pool with a number six Black Doctor fly early in the morning. The Penobscot Salmon Club purchased the fish and shipped it the same day to President Eisenhower.

Two weeks later, Maine's congressional delegation formally presented the fish on a huge silver platter. Wearing a broad grin, Ike talked about Maine and fishing.

"You know, the only time I've ever caught a salmon was in Canada," he told the lawmakers while photographers crowded around. "By golly, this is a nice-sized one."

"Should be good eating, too," remarked Maine Senator Frederick Payne.

The president agreed; he liked fish and he liked fishing. He fished the Mud Creek near Abilene as a boy, and later the nearby Smokey Hill River and Lyons Creek. When he was in Europe, Eisenhower escaped the war for a few days to go salmon fishing in Norway.

Broiled fillet of trout was one of his favorite meals, and Mamie Eisenhower often served dishes made from canned salmon; she had her own recipe for salmon loaf. At a spring luncheon for Senate wives, she served salmon mousse while a recording of bird songs played in the background.

President Eisenhower told the delegation he had never been to Maine and explained the kind of fishing trip he'd like to make. The Secret Service, however, did not think it a good idea for the president to travel in a tiny plane to some remote part of the North Woods. And besides, Eisenhower needed to prepare for Winston Churchill's upcoming visit to Washington.

Churchill hoped to harmonize the two nations' conflicting policies on coping with Communist aggression. Eisenhower hoped for new progress in his drive to enlist Britain and other Western countries in a "united front" to block Red expansion in Southeast Asia.

International issues dominated Eisenhower's agenda. Domestic issues, such as energy policy and water issues, he left to states and towns to address. He vetoed funding for municipal wastewater treatment plants, and prevented further legislation for more federal support of local pollution control facilities. Eisenhower wanted free markets and private enterprise to create the growth required by an emerging consumer lifestyle. Growth was good, and growth required energy and water. Therefore, dams and other developments continued to be a public good. Seeking to make a definitive break with the Democratic past, Eisenhower's first State of the Union address indicated his administration would favor private hydroelectric power development over federal dams.

Like Truman, Eisenhower favored the Colorado River Storage Project, which included the Echo Park Dam in Dinosaur National Monument, a protected area that President Franklin Roosevelt had expanded by 200,000 acres. The proposed dam had ignited a national debate over wilderness preservation.

With conservation interests focused on public lands in the West, the New England-New York Inter-Agency Committee went about its work assessing natural resources in the Northeast in relative quiet.

As part of the 1950 Flood Control Act, Harry Truman had directed his natural resource agencies to initiate a comprehensive survey of the lands and waters of twenty-eight river basins in the Northeast, and to make recommendations for their development, use, and conservation. The resulting study and report identified sixteen out of more than one hundred possible locations in the Penobscot as suitable for new hydroelectric dams: five sites on the East Branch, three on the West Branch, one on the Mattawamkeag River at Stratton Rips, one on the Piscataquis River at Maxfield, four on the mainstem Penobscot, and one on a proposed "Penobscot River Diversion at Bangor." The "economically marginal" diversion plan would build a new dam across the main river channel below Sunkhaze Stream and just above the existing Milford Dam. Water from the new reservoir would be diverted via canal twelve miles west to Pushaw Lake. Another canal at the foot of Pushaw would then shunt the water five miles through Caribou Bog and Penjajawoc Marsh to the Bangor Dam. Building all of them would increase the total amount of storage by more than one million acre-feet at a total cost of $221 million.

In considering how the plan would affect fish and wildlife resources, the committee reported that "Atlantic salmon were formerly present in the streams of the watershed. A few are still taken annually at Bangor, but the annual run is now very small as compared to its former size. Dams and pollution have caused this decline and have made the limited plantings of young salmon ineffective in rebuilding runs of fish."

The three dams proposed for the thirty-six miles between Ripogenus Dam and North Twin Dam would flood the most valuable salmon spawning areas on the West Branch Penobscot River. The five East Branch dams would inundate thirty-four miles of prime canoeing and fishing waters and eliminate all opportunity for restoring Atlantic salmon to the watershed. Construction of more dams on the mainstem would have a cumulative negative effect on the passage of salmon moving upstream and downstream, and eliminate the small amount of remaining salmon nursery and spawning areas, and would strain a river already struggling to absorb pollution.

The committee acknowledged that if the complete inventory power plan for the Penobscot were constructed, there would be no valid reason for the inclusion of fishways at dams, because restoring Atlantic salmon to the river system would be impossible.

The committee insisted that their multi-volume "Coordinated Basin Plan for the Penobscot River" was only an inventory, not an authorization of particular projects. By the time it was released, conservationists had defeated the Echo Park dam proposal and were working furiously on national wilderness legislation. No one made a serious attempt to implement the findings of a government committee planning dams for the Maine woods.

Private interests had the Penobscot dams well under their own control. Great Northern Paper completed construction of a $7 million hydroelectric station at Ripogenus Dam, the company's fifth and largest hydropower dam. Instead of flowing through the upper gorge, water was diverted into a tunnel that ran 3,850 feet from the dam to the power plant. They also expanded the East Millinocket mill, adding two new machines at a cost of $38 million.

In 1950s post-war America, progress had costs, and Americans accepted those costs, because to find some other way to be in the world, a way that would allow salmon to run up the river like molten silver to be caught for food and spirit, required a sacrifice too many were unwilling to make or unable to imagine. Power demand was increasing, they said. Build more dams—fish ladders can carry the salmon. Leave the mills alone—the river will dilute their waste. The Penobscot reflected this form of progress—pollution became the dominant force in the river.

The anglers continued to meet at the Penobscot Salmon Club, and to advocate for restoration of the river. Former commercial fishermen were still around who remembered how to tend a salmon weir. And the region's native

Wabanaki people held salmon somewhere deep in their cultural memory. But for everyone else, the bounty that once filled their backyard river was forgotten.

Eisenhower never got another Penobscot salmon. Salmon numbers dwindled so low that the Atlantic Salmon Commission shut down the recreational fishery. The salmon pool emptied, and the presidential salmon tradition came to an end.

Chapter Eight

Penobscot

The East and West Branches of the Penobscot River meet at Medway, forming the mainstem or "lower" Penobscot River, but not before the West Branch crests one final dam, the Rockabema or Medway Dam. With more than half of its ninety-eight miles impounded by dams and diverted through hydroelectric stations, the West Branch contains the largest portion of inaccessible salmon territory in the Penobscot watershed.

Below Medway, the river flows wide and dark with so much forest water. Out of the mountains, the river begins to meander, gradually, taking on the familiar sinuous appearance of other large waterways with intact floodplains. The outside of the curve is always fastest, deepest (the thalweg); the inside shallow and swampy. The river flows even more slowly because not too far below Medway is the Mattaseunk or Weldon Dam, built in 1939.

Below, the river flows unimpeded for twelve miles, and picks up the Mattawamkeag River, an undammed tributary that flows fifty miles from the Canadian border through extensive, flat bogs and fens, and industrial forestland. The tributary has worn through soft green slate, made potholes in bedrock ledges. In the woods, erratic boulders of Katahdin granite testify to the glacier that formed this place. Some of the tributaries, like the East Branch, Beaver Brook, and Big Gordon Brook, provide high-quality salmon spawning habitat.

Below Mattawamkeag, the channel begins to divide and braid around islands; the banks are low and forested, with roads running parallel to both sides of the river. After flowing over the West Enfield Dam, the Penobscot River is met by the Piscataquis River, the last major tributary.

Up the Piscataquis, beyond the mill towns of Milo, Dover-Foxcroft, and Gilford, prime Atlantic salmon spawning habitat can be found along the upper river. The Pleasant River flows through the waterfalls of Gulf Hagas, "the Grand Canyon of the East." The river falls hundreds of feet in just four miles, through folded mudstone and between high vertical cliffs. The Sebec River drains Sebec Lake, one of the original landlocked salmon basins.

All of this energy flows into the main channel of the Penobscot.

The wide, shallow river collects water, light, and heat, flowing south between green shorelines interrupted only by a few modest homes. Fishermen cast for smallmouth bass, a non-migrating member of the sunfish family introduced to the Penobscot in 1869; salmon co-exist with bass in freshwater, as well as brook trout, blueback trout, yellow perch, chain pickerel, cusk, hornpout, sunfish, white sucker, sticklebacks, chub, dace, shiners, minnows, and whitefish. Atlantic salmon struggled along, extinction hovering like a cloud of just-hatched mayflies, and bass became the top game fish in the river, likely the first fish caught by children who were told not to eat the fish, not to swim in the river.

Rivers are all about flow. Fast, narrow brooks tumble out of the forested hills and down the sides of Katahdin, curving around rocks and trees, swaying from shallow riffle to deep pools, merging into the wide, deep channel of the lower mainstem. As the river moves, the surface layer pulls on the water below, and the bottom layers pull on the river bottom. The river bounces, falls, and drags down to the sea.

Everything in a river is adapted to this flow, and the habitat the flow creates, from the film of photosynthesizing algae, fungi, and bacteria that coats rocks, to the insect larvae that graze on the slimy film, and bugs that chew on moldy fallen leaves. Bits of algae scraped from rocks, leaves shredded by larval stoneflies and mayflies, drift downstream and are collected and filtered by other hungry stream dwellers that are in turn eaten by salmon. Energy, nutrients, spiral in the flow, spiral into the sea, into salmon.

Flow created the Atlantic salmon's physique, and formed its identity as The Leaper. What happens when there's nothing to leap, no flow to follow?

Dams violate the river continuum. Dams interrupt the flow, destroy the drag, flood the riffles. Each dam is a fracture, and eventually the river breaks.

Cardinal flowers splatter the wet banks with splotches of brilliant red, giving way to purple asters and meadowsweet in the fall. Far from the rugged headwaters, and the flooded lakes of the North Woods, the lower Penobscot River is a series of reservoirs, one after another after another, a sluggish sheet of water trying to be a river.

Beneath the surface are smolts trying to be salmon.

In their second or third year, when they have grown longer than four inches, young salmon parr begin a major transformation that prepares them for life at sea. Their bright colors and dark banded camouflage turn to silver as light-reflecting spicules of guanine and hypoxanthine deposit in the skin and scales. This silvering, common to many fish that live near the surface, presumably makes it harder for predators to see them from above. The fork in the tail deepens. The body elongates, slims, streamlines. The fish turn around to face downstream.

As the days lengthen and the river warms, typically in early May, changes happen inside the fish, too: it becomes more buoyant and starts producing an enzyme that will help it cope with salt.

In water, salt ions seek equilibrium, moving from areas of high concentration to areas of lower concentration through osmosis. All living things need salt to live, and so freshwater fish work to maintain their internal salt content by pumping sodium and potassium into their bodies through specialized enzymes in their gills. But when salmon become ocean dwellers, the situation reverses: now they have to keep too much salt *out* of their bodies. Their gills change physically and chemically. The transition takes energy, and many salmon fail to cope, but some of the territorial and precocious parr become schooling silver smolts.

Within weeks, river flows increase and temperatures warm, and smolts begin to migrate downstream with their siblings to the sea. They take their time swimming down the Penobscot River, moving at night and hunkering down during the day. The closer they get to tidewater, the faster they travel.

As damaging as dams are for adult salmon migrating upstream, converting a flowing river to a series of slackwater impoundments is even more problematic for the young fish attempting to reach the ocean. The route that a salmon smolt takes when passing a dam is a major factor in its likelihood of survival.

Dams alter water flows and temperatures, both of which affect the smolting process and the timing of migration, potentially delivering the smolt to the ocean at the wrong time. Smolts that are delayed while attempting to find a way downriver are snatched up by predators like bald eagles and cormorants, or suffer when water warms in summer. A fish that passes through a properly designed downstream bypass has a better chance of survival than a fish that goes over a spillway, which, in turn, has a better chance of survival than a fish swimming through the turbines.

When passing over (or through) a dam, downstream migrating smolts won't survive a free-fall that is too high or too fast. They are killed in turbine blades, or die from gas bubble disease when too much air gets trapped in turbulent water. Injuries can lead to infection and slow death. For a long time, biologists and engineers assumed that 10 percent of smolts would die by these various means *at each dam* during the downstream migration.

The channel braids around numerous islands: from small, sandy deposits fringed in pickerelweed and ferns to wooded groves where sunlight breaks through great arched galleries of silver maple, ash, and birch. In the shallows, in between patches of bur-reed, arrowhead, and bulrushes, exposed sand is stitched with tiny animal tracks. Freshwater mussels cover the bottom, peeling bronze coins half-buried in mud.

> Fiddler Island: bald eagle flying from a white pine.
> Grass Island: great blue heron stalking fish, Canada geese honking beneath two eagles high overhead.
> Olamon Island: place of red ochre, powdered hematite, *Olamon*, the people's red paint.
> Sugar Island: covered in maples, source of sap, syrup, sugar.
> Freese Island: worn pyramids of rock and wood, old logging boom rising out of the river, boom crumbling into the river, boom slowly eroding, fading.

The islands are part of the Penobscot Indian Nation's territory, which extends from Indian Island to the more than 200 islands upstream. No Trespassing signs posted on the islands are less a warning than a proclamation. *We are here. This is our home. Where do you come from?*

In the 1818 treaty with Massachsuetts, the Penobscots retained their rights to the islands in the Penobscot River. In 1824, the Maine Legislature passed a resolution authorizing the governor to negotiate with the Penobscots to sell their islands to the state. Nicatow Island, at the fork in the river where it splits into the East and West Branches, sold in 1828. The Penobscots held on to the other islands, living on some and using others as base camps for fishing and hunting or leasing the timber to logging companies.

At the federal level, attitudes (and thus policies) toward Native Americans shifted with each administration, from sympathetic awareness under Roosevelt's "Indian New Deal" to assimilationist or terminationist. But Maine tribes were not recognized by the federal government and thus were not subject to federal policies. The Department of Health and Welfare (not the Bureau of Indian Affairs) was "responsible" for Maine's Indians, including their education (off-reservation), food (through federal surplus food programs), and (sometimes) income, provided from a trust fund established by the Maine Legislature and funded by tribal land sales and leasing. Indian agents, the state workers administering state welfare policy, were untrained and abusive. They felt the tribes asked for too much, lacked initiative, and they questioned the legality of requests for medical care, voting rights, and fees from those who used their islands.

In the 1950s, the state felt the tribe might be better off selling the islands in the Penobscot River.

The Maine Department of Health and Welfare suggested that the state purchase the islands at a standard but "fairly generous" rate. The Forestry Department would then administer the islands as the "Penobscot Indian Forest" and allow tribal members to erect temporary camps on the islands for fishing and hunting—after securing a permit from the Forest Commissioner.

To establish the real estate value of the islands, the state had them surveyed. The James W. Sewall Company's report found 146 islands between Old Town and Mattawamkeag. The islands had changed in number and shape over the years under the influence of the river's current and flooding by dams. The report focused on the trees and the lumber they could provide. Hog Island had "poor hardwoods;" Island #85, a few poor hardwoods; Island #88, nothing; Island #95, generally poor quality, low land hardwoods; Hockamock Island, some worthless swamp; Mink Island, only a few trees. The surveyors struggled to align the islands with their maps without deeds to trace or boundaries to follow.

With the tree growth on the islands being so "meager," the state figured the Indians would be eager to dispose of their lands. "The suggestion that a Penobscot Indian Forest be created is not being made with any thought of trying to deprive the members of the tribe of lands," wrote David Stevens of the Department of Health and Welfare. "It is believed that as long as the present circumstances continue—that is, the uncertainty as to title of the lots involved and the impossibility of identifying the lots on the face of the earth—that the Department of Health and Welfare must refuse to approve leases for stumpage. For this reason, the Indians will not derive any revenue from the lands and there will be no use made of these lands." To re-survey the lots, or figure out the titles to the land to enable proper assessment of timber value, would "cost more than the lands themselves are worth."

The Penobscots declined to transfer the islands.

Native resistance to such proposals had been continuous, but gained momentum in the middle of the twentieth century. When young Penobscot veterans returned from World War II, they carried with them knowledge of the world beyond their homeland, and a stronger interest in their own culture and heritage. Some left Maine to find work elsewhere, but the ones who stayed became leaders, working to document how their rights had been undermined.

Discussions of historic treaties helped to unify the Wabanaki when state and federal governments accelerated efforts to terminate Native American tribes and abolish their reservations, such as by selling the islands.

Showing defiance, imagination, and courage, tribal leaders in Maine restored their governing body. They used the media to make their unheeded concerns public. They hired lawyers, petitioned and lobbied the legislature, Congress, and the United Nations, and recruited non-Indian supporters. Reform efforts during the 1950s, while largely unsuccessful, brought a growing awareness among tribal leaders of the need for political autonomy, including jurisdiction over tribal lands and rights to hunt and fish.

In 1952, Penobscot Governor Albert Nicola declared that the Penobscots, still without voting representation in the state legislature or control of their own territory, were being dictated to. He encouraged tribal members to unite in resistance.

The ledges begin at Orson Island. When the last ice age ended, the river meandered east and west, using different channels through layers of glacial sediment and old ocean floor, before finding its current course. In many sections, topography forced the river to flow over freshly exposed bedrock, resulting in waterfalls and whitewater rips. The numerous rapids and shallow falls in the twelve miles of river between Old Town and Bangor gave rise to the name *Penobscot*—place where the river tumbles over diamond bedrock, waters of descending ledge where the rocks widen out. The bedrock is five-hundred-million-year-old metamorphic and sedimentary rocks, soft shale, slate, and schist—rocks that "stand on their edges" as Thoreau wrote.

The same ledges that form the staircase rapids of the mainstem Penobscot River, that gave the river its name and created prime fishing grounds, made ideal foundations for sawmills and dams. Old Town Falls was the site of the first dam that stretched all the way across the river channel in the 1820s. The Milford Dam and its 1905 companion dam at Gilman Falls on the western channel of the Penobscot (known as the "Stillwater River") flooded 235 acres upstream. The three-mile impoundment encompassed several islands, including all of Indian Island. After futile protest, the Penobscots watched the dam flood out their best fishing grounds, the falls where every year for thousands of years silver salmon were caught as they surged upriver.

In the 1950s, this stretch of river was covered with floating wood chips, mats of stinking sludge, water cast gray by a white bacterial fuzz covering the river's edges and bottom. Like so many rivers across America in the middle of the twentieth century, the Penobscot had been turned into nothing more than a convenient conduit to carry the worst of what was thrown away.

Untreated human sewage flowed from towns and cities along the river. Tanneries ejected greasy, acidic remains of rendered flesh, chromium, and caustic chemicals. Shoe factories, mills, meat packing plants, bottling plants, and dairies all contributed waste to the Penobscot. Fibers and toxic chlorobenzene dyed the water below woolen mills. Industrial wastes had been dumped in the Penobscot since sawmills began operations in the late eighteenth century.

In 1867, Maine fish commissioners Foster and Atkins noted that great drifts of sawdust "settle down upon river bottoms that were before well peopled with insects and other small creatures, and destroy all life." Mill waste blocked ship traffic to such an extent that the government had to dredge it out, while the individual mill owners had long and bitter lawsuits over it. Rufus Dwinel successfully sued lumber baron General Samuel Veazie, claiming that waste from Veazie's mill interfered with his mill upstream.

An 1868 "Edging Law" prohibited disposal of slabs, boards, bark, grindings, or lath edgings into the river, and the 1899 Refuse Act required a permit from the Army Corps of Engineers to dispose of solid material, but only in navigable waters (below Bangor in the Penobscot). Sawdust still drifted downstream, and accumulated enough to block boat traffic and clog fishing nets. Fine, lightweight sawdust particles settled out in the estuary as the tides mixed back and forth, forming a soft carpet of decomposing wood particles, up to two feet thick in some tidal flats.

"The extensive deposits have in some instances so altered the configuration of the bottom as to interfere with the success of certain fishing-stations," Charles Atkins noted, "but beyond that I see no evidence that the discharge of the mill refuse into the river has had any injurious effect on the salmon. It does not seem to deter them from ascending, and, being thrown in below all the spawning grounds, it cannot affect the latter." Still, the following year he demanded "a more rigid and faithful enforcement" of the law forbidding the throwing of mill waste into the Penobscot below Medway. Biologist Richard Cutting noted that sawdust and wood waste was still being dumped into the river in 1959.

None of this waste could match the waste of the pulp and paper mills, which accounted for most of the pollution in the Penobscot River.

The Old Town paper mill, constructed in 1882 by the Penobscot Chemical Fibre Company at Great Works, was the first chemical pulp mill on the river; the South Brewer mill was the first sulphite mill.

Manufacturing pulp required tremendous volumes of water: millions of gallons each day to strip bark from logs, or wash pulp from grindstones. Mills used different methods to break down the lignin that binds the long cellulose fibers of spruce and fir. In sulphite mills, an acid sulphite solution digested knotted, split, and chipped wood under steam pressure and high temperatures. Others used an alkaline digestion process that produced lime sludge and carbon or "black ash" from incinerating spent pulping liquor.

Then more water was required to wash, screen, and bleach the fibers before they were washed again, dried, beaten, and flattened into paper. Water to carry much of what remained—black ash, sludge, and spent pulping liquor, cloudy with fiber fillers and sediment—into the river.

The impact of the paper industry was visible within a decade of its inception.

In 1890, sportsmen feared that "already the extensive pulp mills on the river above—with its many branches also beginning to be lined with pulp mills—are about to show their deadly work upon the salmon of the Penobscot. The chemicals that are discharged into these streams are believed to be death-dealing to the salmon. Last year it was particularly noted that the ascending salmon invariably crossed over to the other shore, where it was possible to avoid the deadly chemicals, and that frequently they were found dead."

U.S. Bureau of Fisheries naturalist William Converse Kendall visited the Penobscot in 1904. At the junction of the East and West Branches, seven miles downstream from the Great Northern paper mills, he found the bottom everywhere covered with waste pulp. Where the river flooded high in spring, the shrubs and bushes along the banks were swathed in ragged shreds, brittle and dry like a hornet's nest.

Maine led the nation in wood pulp production.

In 1917, the Maine Legislature passed a law prohibiting the discharge of sewage and other polluting waste into public water supplies, but exempted the Penobscot, Kennebec, Androscoggin, and Saco Rivers. An amendment in 1929 to add pulp and paper mill effluent, including process water, and limit the exemption to tidewater, was defeated. But the paper industry set up a committee, chaired by Joseph Warren, and provided $10,000 to study the effect of their discharge on dissolved oxygen in the rivers. Pulp and paper wastewater contained organic matter that used up oxygen as it decomposed. For example, cellulose liquor from the Eastern Manufacturing Company at South Brewer demanded 900 parts per million dissolved oxygen in the course of a day.

The study team found "marked reduction in the dissolved oxygen content of the water from above Millinocket to just above the junction of the East Branch," which they attributed in part to the large cordage of pulp wood stored in the impoundments above Quakish and Dolby dams. The committee of paper industry representatives wrote that "it has long been accepted—and still is today—that our streams and rivers must serve as natural channels for the carrying off and removal of liquid waste materials . . . "

At the time, health officials and engineers considered three parts oxygen per million parts water the minimum for supporting fish. Biologists later increased the lower oxygen limit for cold-water fish like salmon to five parts per million. When oxygen levels drop suddenly, salmon go to the surface, piping for air, and then die, gillflaps flared, mouths agape. Where low oxygen levels persist, death comes slowly. Fish stop eating, become sluggish and vulnerable to predation and infection.

Along the lower West Branch, dissolved oxygen levels fell below five parts per million. "With the exception of that portion . . . the river appears to be in good condition and to be successfully receiving such waste as is at present introduced into it." Decomposing waste also led to acidic conditions; low pH was detected below Millinocket, East Millinocket, and Great Works.

Despite the study's rosy conclusion that "the weekly averages of the dissolved oxygen content of the water of the rivers at practically all places was capable of supporting fish life," a new and tougher amendment was introduced in 1931. Again the paper mills fought to keep their exemption, claiming the restrictions would handicap their industry.

The Department of Sea and Shore Fisheries, in their 1936 report, requested better cooperation from mill owners in eliminating pollution. But the pulp mills were on track to triple their daily plant capacity, and Maine would lead the nation in paper production in the 1940s and 1950s.

Decomposing organic matter became suspended in the water, causing heavy turbidity. Rank odor, unsightly scums, waste coating the stream bottoms—the pollution was a blanket smothering the river. Bacteria fed on the waste dumped by the mills, building up sludge and using up oxygen. By the 1960s, dissolved oxygen levels reached zero at some locations.

Industrial waste combined with untreated sewage created hostile conditions in the estuary below Bangor, where pollution most grossly revealed itself. Most of the pollution in the bay came from upstream sources, but poultry plants in the local towns contributed their share. Chicken entrails floated in the harbors and a film of animal fat and feathers coated the water.

Blocked from migrating beyond Bangor by low oxygen, some salmon died, some took refuge in Cove Brook and other tributaries of the estuary. Others left, never to return. Some salmon trying to get back to their natal stream had imprinted on polluted water—pulp and paper mill effluent, not pine and slate, was the scent of home. With pulp and paper mills on all of Maine's major rivers, they all smelled the same to salmon. Fish tagged in the Penobscot were found in the Kennebec—they smelled the Kennebec River paper mills and thought they were home.

"Today's Penobscot, with discolored waters and carrying scum and exhaling offensive odors, repels sportsmen and other recreationists," noted Richard E. Griffith, regional director for sport fisheries and wildlife with the U.S. Department of the Interior, at a 1967 conference on Penobscot Bay pollution. "Downstream from Bangor, the Penobscot is so severely polluted that boats cannot be kept in the river because of the way the river fouls the paint."

At the federal level, Franklin Roosevelt's administration had been the first to address water pollution in a serious way. Congress passed the first inland water pollution control bill in 1938, establishing a division of water pollution in the Public Health Service and providing grants to states and towns for sewage treatment facilities. The Federal Water Pollution Control Act, passed under Truman's watch in 1948, provided the first federal funds for state water pollution control programs and sewage treatment plant construction. The act was renewed in 1956 with more solid funding for sewage treatment plants, but without enforceable laws or standards. Pollution was portrayed as wasteful; by making water cleaner, it could be reused by industry. Dams were part of the solution: more dams meant more reservoir storage, and thus more fresh water to dilute the wastes that threatened economic development.

In the years after President Eisenhower received his last salmon, the Penobscot River had become so polluted that no one wanted to go near the river, let alone wet a line in it. There was no opening-day breakfast at the Penobscot Salmon Club, no competition for the first fish, no glory for delivering it to the president. No one was racing to catch the Maine salmon requested by President John F. Kennedy for his Fourth of July dinner in Hyannisport in 1961.

Most anglers either gave up or traveled east to coastal rivers where conditions were better. The Down East watersheds of the Narraguagus, Pleasant, Machias, East Machias, and Dennys Rivers remained hospitable to salmon because they were sparsely populated, heavily forested, and had fewer dams. The Down East salmon rivers flow through bogs of spruce and tamarack and shallow, sandy blueberry barrens, and empty into the ocean in a series of small rocky bays. The Narraguagus River at Cherryfield, not the Penobscot at Bangor, became the frequent site of the first Maine salmon caught each year. For the decade from 1955 until about 1966, biologists with the Atlantic Sea-Run Salmon Commission had no knowledge of salmon in the Penobscot at all.

Even while they pushed for restoration, Maine's salmon biologists questioned whether it would work. "Restoration of the Atlantic salmon to the Penobscot River depends on the installation of efficient fishways, upon a pollution abatement program particularly for the main river, and upon the recognition of minimum flow agreements in regulated waters," reported Richard Cutting. "In turn, these three phases of the restoration program depend on the interest, desire, and initiative of the people of Maine to preserve this fading resource. Public indifference will doom the Penobscot River Atlantic salmon resource, and the Penobscot River will be lost for all recreational uses."

The fishery moved east, and the Penobscot River salmon clubs lost members and the details of angling season drifted from the newspaper's front page to the sports section. Fly rods collected dust; streamer patterns faded from memory.

Some salmon stocked in the headwaters found their way downstream, navigated through suffocating reaches of putrid filth to the sanctuary of the subarctic ocean, with no knowledge of what would be waiting for them when they returned, or if they would be able to get back home.

Chapter Nine

Troubled Waters

Saturday, May 9, 1964

When Cherryfield resident Harry Davis caught the first Atlantic salmon of the year in the Narraguagus River, everyone was ready. State salmon biologists had heard that anglers wanted to revive the presidential salmon tradition, and they were eager to help. They knew that if restoration programs were to succeed, they needed the support of recreational anglers. And so they planned for the first fish of 1964.

The Atlantic Sea-Run Salmon Commission took the fish to H.P. Hood and Sons, a dairy plant with refrigeration know-how, who packed the fourteen-pound fish in dry ice and shipped it via Air Express to Washington, where the Maine Department of Economic Development had arranged for the presentation to President Lyndon Baines Johnson.

Maine's Republican Congressman Clifford McIntyre arrived at the lawn on the south side of the White House to present the gift to staff assistant John McNally. Representative McIntyre had hoped the president would receive the salmon personally, but he was told it couldn't be arranged. Or maybe Johnson didn't want to grant the Republicans a photo op or a chance for a headline in a precarious election year.

After becoming first family in the dark days that followed the assassination of President Kennedy, the Johnsons had imbued the White House with southern charm, optimism, and a commitment to carry on the Democratic policies of the Kennedy years. In the spring of 1964, the Johnsons had not yet replaced Rene Verdon, the French chef hired by Jaqueline Kennedy.

Verdon had chafed when he saw what he was up against: a president whose favorite dish was chili con queso, who preferred steak, chicken, even jalapeno hash over poached salmon. President Johnson liked to fish and hunt on his Texas ranch, but once in Washington politics he scarcely noticed the food he ate at banquets or state dinners. He liked food he could eat quickly, on the run or while in conversation. Simplicity ruled and steak reigned supreme. So maybe the family cook, Zephyr Wright, brought in by LBJ from Marshall, Texas, cooked the salmon, but most likely Verdon served it in a more formal manner.

Chef Verdon sliced the salmon into steaks. In a large skillet, he sauteed chopped onion and celery in butter, then added water, vinegar, bay leaf, thyme, and seasonings, and simmered the broth for five minutes. He wrapped the salmon steaks in cheesecloth, placed them in the pan, covered it, and turned down the flame to gently poach the fish.

Salmon fishermen renewed the Penobscot River "first fish" tradition amid a groundswell of public concern for the environment. Anglers and hunters found their constituency expanded by hikers, paddlers, campers, and back-to-the-landers. Debates about the wisdom of preserving undeveloped lands outlasted controversies over the Echo Park and Grand Canyon dams—and were fueled by regret that Glen Canyon had been sacrificed in the process. Advocates, supported by an exponential increase in state and national conservation organizations, pushed for a national system of wilderness areas and wild rivers. Rachel Carson's *Silent Spring* created an awareness of chemical hazards. "Ecology" and "environment" were becoming common words in the households of a middle class with time and desire to enjoy the outdoors, and new questions about American values and institutions. They wanted wilderness in the West, whether or not they ever visited it. At home, they demanded safety, security, clean air, and clean water.

A few months after receiving his Maine Atlantic salmon, President Johnson signed the Wilderness Act, protecting nine million acres of federal land. Johnson knew that conservation had to move beyond protecting remaining wilderness to restoring what had been lost. His view of "environmental" issues went beyond wilderness to include urban problems such as noise, overcrowding, and pollution. His kind of conservation provided recreational opportunities for all Americans, regardless of where they lived. The Atlantic salmon clubs and their fishing pools, in close proximity to towns and cities, fit well within Johnson's vision for a Great Society and a beautified America. In his "Message to Congress on Natural Beauty," President Johnson spoke of uplifting the dignity of the human spirit by turning the federal government's attention to improving the relationship between people and the world around them. He listened when his supporters demanded government action on water pollution.

Within a year of Eisenhower's veto of federal funds for wastewater treatment plants, President Kennedy signed legislation creating a $100 million annual subsidy program. But the federal government still had no enforceable standards for water quality, and the House of Representatives opposed federally mandated pollution limits.

On Capitol Hill, Maine Senator Edmund Muskie and his colleagues on the Senate Public Works Committee were reconstructing federal water pollution laws, with the support of the Johnson administration. Environmental activists in Maine, who invited Muskie to be a keynote speaker at a Portland forum ("At What Price Clean Water?"), had convinced him that clean water was an important issue worthy of his attention. Muskie drew upon his childhood next to the Androscoggin River, and his experience with the S.D. Warren Paper Company and other manufacturers who repeatedly tried, and failed, to predict the "assimilative" capacity of Maine's waterways in lieu of treating mill wastewater.

Not until the overdose of waste converted large stretches of rivers into sulfur-spewing sewers did the chemists and engineers admit the uncertainties inherent in their methods. Muskie called for a national program to prevent water pollution at its source, rather than attempting to cure pollution after it occurred. Water quality standards should recognize a stream's best potential, rather than simply "locking in" present conditions.

The Public Works Committee produced a film to support their work. In *Troubled Waters,* narrator Henry Fonda patiently and firmly explained that "if America's waters are troubled, it is because they are overworked. Water can purify itself, if there's enough time and space between the jobs men ask it to do." Water was a servant to be employed and polluted, but not so much that the next stranger downstream couldn't do what he needed with it. Water was a limited resource to be stretched to its limits, with the help of treatment plants that sent rivers downstream, "ready to do their job."

The film features Muskie's home river of the Androscoggin, where chemicals "strong enough to turn solid wood into paper," turned the water brown all the way to the sea. "The Indians called this 'river of many fishes,'" informed Fonda; "They wouldn't call it that now, and they wouldn't eat the few fish they could catch."

Still strangers in their own country, Americans assumed everything around them would absorb their excess. America, in pursuit of energy, of wealth, had reached a moment of ecological poverty: acidic rain, ruined rivers, and empty nets. If ever there was a president who hated poverty, it was Lyndon Johnson.

In his pledge of "Clean Water by 1975," President Johnson condemned pollution in the Potomac River. No longer a local problem, as Eisenhower had viewed it, pollution was a nationwide blight that demanded solutions from scientists and engineers. Rivers catching fire, water undrinkable, fish poisoned to death and washing ashore. Pollution had forced the closure of one-fifth of the nation's shellfish beds. More than 90 percent of America's waters were polluted.

Johnson stopped in Portland, Maine, on the campaign trail in September 1964, a few months after he had received (or not received) his salmon. Governor Reed had proposed a $25 million bond for water pollution control, to cover the state's share of an estimated $97 million needed to clean up Maine's waterways.

Had he visited Bangor, President Johnson would have learned that the filthy Penobscot, little more than "an open sewer," "a wasteland," needed $35 million alone in remediation to meet minimum water quality standards that would sustain Atlantic salmon. Some 150,000 people lived in the Penobscot River watershed, but the pollution load was equivalent to a population of five million.

Pollution forced Maine's Commissioner of Sea and Shore Fisheries to close nearly all the remaining clam flats in Penobscot Bay, a fishery worth $5 million. The Federal Water Pollution Control Administration investigated, motivating industries, municipalities, and fisheries managers to finally address pollution in the Penobscot.

The Maine Water Improvement Commission was preparing recommendations to the legislature on how to classify segments of the Penobscot River, from "pristine" AA waters to "nuisance" Class D, based on how much pollution the river could handle and how much it was already polluted.

The commission determined that in the Penobscot River, "the paramount pollutant was putrescible organic matter," and classified much of the river as Class D, especially below Great Northern's paper mills and in the estuary. State officials had an "attitude of special tolerance" toward the paper companies, in part because industry lobbyists had great influence in the legislature. They knew the lower classification would pass easily. As long as odor did not become a nuisance, the river was to assimilate as much pollution as possible. Only when conditions deteriorated, when the putrid sludge floated to the surface, was the waste load to ease.

At a public hearing on the classifications, Atlantic Sea-Run Salmon Commission chair Horace Bond urged the water officials to aim higher. Bond was not a fisheries scientist. He was a grain salesman, an angler, and an inventor of several fly patterns, including the Silver Salmon and the Songo Smelt. But Horace Bond was persistent. He took the job with the salmon commission in part because he believed the river could come back.

Pulp and paper industry representatives, who far outnumbered salmon advocates at the hearing, asked for reason. They reminded the commission how much they'd already spent on pollution control, how cooperative they had been, how much they contributed to the state's economy, how they would have to move south if such restrictive measures were enforced, leaving thousands of unemployed Mainers in their wake.

The state upheld the Class D rating.

When the federal government, which had gained the power to review state water quality standards, objected, the state reluctantly reclassified most of the river to Class C ("water suitable for fish life without scums or odors with a dissolved oxygen content of at least five parts per million"), but kept the lower standard for the stretch of river below Great Northern Paper. The cover of the classification report featured a line drawing of a fishing tackle basket and fly rod.

Even with the lower classification, Great Northern had to reduce waste by at least half just to meet the Class D standards. And the state's approved cleanup schedule gave the pulp and paper mills nine years to adjust processing and add treatment. The delayed timeline would mean many more years of pollution.

Muskie's hard work culminated in October 1965 with congressional approval of the Water Quality Act, the most important legislation of its kind during the sixties. But the real cleanup would have to wait; for the rest of the decade, federal attention was turned to the jungles of Southeast Asia.

With pollution-choked dead zones and inadequate passage at the dams, only the very strongest of salmon were able to reach spawning beds upriver. Since 1872, the Craig Brook fish hatchery had put 56 million salmon into the Penobscot River, but by 1964 scientific studies had confirmed, again, that Atlantic salmon runs could not be restored or even maintained by stocking alone.

Americans had come to value not just wild landscapes, but the plants and animals that inhabited them. The nine biologists who composed the Committee on Rare and Endangered Wildlife Species, established within the Bureau of Sport Fisheries and Wildlife, included Atlantic salmon on their preliminary list of wildlife thought to be in danger of extinction in 1964. In 1966, Congress passed the Endangered Species Preservation Act in an attempt to address species endangerment in a more comprehensive way. The first official list of threatened and endangered species included shortnose sturgeon, a migratory fish native to the Penobscot River, but not salmon.

Still, Maine's salmon finally had powerful advocates in federal government, including Muskie and Secretary of Interior Stuart Udall. A lawyer and son of an Arizona Supreme Court justice, Udall was appalled by waste and America's post-war preference for personal satisfactions over public needs. He called on Americans to challenge the assumptions that were "braided into a chain of myths and priorities" and led to a "domestic willingness to settle for materialistic mediocrity." He saw so much waste around him that could have been solved with coordinated planning and technological tools of renewal.

Scientists were Udall's symbols of conservation, and he supported their plans for Atlantic salmon. He directed state and federal agencies to develop a model of restoration, building on the work begun in the 1940s by the U.S. Fish and Wildlife Service and the Atlantic Sea-Run Salmon Commission.

In reviewing the salmon rivers of Maine, the Atlantic Salmon Commission concluded that the Penobscot River offered the best hope for such a restoration plan. To try to restore all the rivers at once was not feasible; but by concentrating their efforts on the Penobscot, fisheries managers hoped to create "a model in overcoming the man-made socio-economic problems that affect fisheries and reduce full river realization throughout the United States." Encouraged by Secretary Udall, the Atlantic Sea-Run Salmon Commission intended to develop measures that could be used to restore other rivers in Maine, New England, and the country.

Maine's fisheries managers hoped the restoration would create a mecca for sport fishermen, and maybe even commercial fishing. They looked forward to the federal grants that would come as a result of the 1965 Anadromous Fish Conservation Act, and the public enlightenment they hoped would help solve pollution problems.

The restoration plan included an estimated $737,000 in fishways to bring back thousands of salmon and allow an annual "sport catch" between 180 and 750 salmon. "If the plan is as good as we expect it to be," said Udall, "we shall be in a position to approve funds for half the cost of the first two fishways at Milford and Great Works."

Like Johnson's vision for greater connections between society and environment, biologists Harry Everhart and Richard Cutting envisioned the Penobscot as a model not just for salmon restoration, but for restoring community connections and offering recreational space close to home for fishing, canoeing, and just enjoying the outdoors. "Restoration of anadromous fishes combined with the pollution abatement program in operation are sure to promote extended recreational use of the Penobscot River," they wrote in their report to the U.S. Fish and Wildlife Service.

In 1967, the Department of the Interior designated the Penobscot a "Model New England Salmon Restoration River." The following year, the Penobscot was one of twenty-seven rivers designated for study as potential additions to the new national system of Wild and Scenic Rivers.

Signed by Johnson in October 1968, the Wild and Scenic Rivers Act followed nearly two decades of controversy over proposed dams within Hells Canyon on Idaho's Snake River. The Department of the Interior sued the Federal Power Commission, asserting the proposed project would have adverse affects on fish and wildlife resources, resulting in a historic Supreme Court decision in which the definition of "public good" was expanded to include environmental values.

Writing for the majority, Justice William O. Douglas challenged the commission to consider whether the best dam might be no dam. The test, he argued, was not solely whether the region would be able to use the additional power, but whether the project was in the public interest, a determination that required an exploration of all relevant issues: future power demand and supply, alternate sources of power, the public interest in preserving reaches of wild rivers and wilderness areas, wildlife protection, and the preservation of anadromous fish for commercial and recreational purposes.

The Act designated eight rivers as initial components: the Clearwater and Salmon in Idaho; the Rogue in Oregon; the Feather in California; the St. Croix in Minnesota; the Wolf River in Michigan; the Rio Grande River in New Mexico; and the Eleven Point River in Missouri. A candidate river, the Penobscot would be subject to further study.

Across the national mall, other federal offices were making their own plans for the Penobscot that would complicate restoration for decades to come.

Undeterred by the Snake River lawsuit, the Federal Power Commission prepared "planning status reports" for major rivers across the country as a first step for licensing hydroelectric dams, as required by the newly amended Federal Power Act. Only one dam on the Penobscot had a license, Penobscot Chemical Fibre Company's plant at Great Works, issued in 1963. License applications were pending or being prepared for the other dams. The appraisal report built upon the work of the New England-New York Interagency Committee a decade earlier. The new planning was to move beyond an inventory of possibilities to "a definite program to be carried out under a definite time schedule." Between 1946 and 1968, Americans' per capita consumption of electricity rose 436 percent. The Office of Science and Technology predicted that 250 "mammoth" power plants would be needed by 1990. Congress had approved the Army Corps of Engineers' plans to study construction of two dams on the St. John River that would flood 140 square miles for 830 megawatts of electricity, most of which would be exported to other New England states.

The Federal Power Commission saw great hydropower potential in the Penobscot. "Substantially greater installations can be justified to meet the growing power demands of the area," their report said. Even though Great Northern Paper's private hydroelectric dams on the West Branch controlled the majority of the storage and power capacity of the entire Penobscot watershed, the commission highlighted three additional possible future projects on the West Branch that would flood what was left of the river below Chesuncook.

The revival of the presidential salmon tradition was brief; in 1965 life in the Penobscot Valley was back to what had become normal: the lingering desire for more dams, more power. Water unfit to drink, clogged flows unfit for salmon, a meager harvest unworthy of the president.

In Orrington, on the eastern shore of the estuary, International Minerals and Chemicals built a factory where workers made chlorine and caustic soda by sending an electrical charge through salt water. The "chlor-alkali" process em ployed mercury as an electrical conductor. International Minerals and Chemicals buried sludge laden with tons of mercury and other chemicals in five unlined landfills on the property, which sat at the top of the Penobscot River's steep banks. And the company was legally permitted to discharge sixteen pounds of mercury to the river every year.

But the rivers were not ruined beyond hope, as was demonstrated in the spring of 1968, when the Eastern Corporation paper mill in Brewer shut down for a few months. Salmon appeared, jumping amid the slime in the pools. Excited, angler Dick Ruhlin got his fly rod and went down to the river. He stood by the river, trying to fish, his eyes watering from the noxious odors. Ruhlin didn't pick up his rod again for five years.

Chapter Ten

Kenduskeag

In the blank days of winter's end, the Penobscot River is sheathed in ice. Wind born on the Siberian plain rattles windows like loose teeth.

In downtown Bangor, Kenduskeag Stream groans with each surge and retreat of the tide, its thick white armor cracking as it heaves against the sides of the canal, ten feet up and ten feet down, twice each day between the concrete and granite walls of the city. Stretching out to the confluence with the Penobscot River, plates of ice break and sink, push together, form ridges and valleys. Solid white globes collect into piles, then slip between the cracks with a thundering splash.

Upriver to the north the ice is three feet thick and black as obsidian, but still within reach of the tides. The Coast Guard's ice cutters gnaw their way to the waterfront then retreat back to Penobscot Bay, leaving a wake of sodden, frazzled ice rafts and floating gray slush. Ice builds up on the edges of the river, slabs stacked upon shelves.

Upriver to the north, golden orange salmon eggs slumber deep in gravel beds; two-year old salmon parr, anxious for their first taste of salt, get ready to smolt. Everyone waits for the sun to strengthen, days to lengthen, ice to thaw, river to flow.

In years past, the good fishing years, the salmon clubs would be stirring, flies being tied, rods fine-tuned, fires stoked as anglers watched the pools, tested the ice with an anxious boot, counted the hours until Opening Day. Salmon, silvered and strong from their winter in the ocean, would be charging through the bay as fast as twenty feet per second, the largest fish traveling up to twenty miles a day, usually under the cover of darkness. Each morning, as the salmon rested, breathed, in pools of water, the newspaper would carry updates on the melt.

By the third week in March, the ice melting fast into rivers and then into bigger rivers, until one day the ice is just gone, the Kenduskeag a torrent of brown and white rapids.

Did salmon fishermen dream about Opening Day? Did they dream about the ice breaking, water flowing, their peapods rocking in the roiling current? Did they wake in the morning with new patterns of feather and floss wrapped in their imagination?

Did the ancient Penobscots dream of rivers turned to silver, food flowing by the ledges at their feet, spears sharpened and waxed?

The name *Kenduskeag* means place of the water-parsnip or eel-weir place, the place for catching eels from the rapids, where now the river is a canal that cuts through downtown Bangor, third-largest city in Maine. The Kenduskeag begins at the outlet of a pond twenty-five miles northwest of Bangor, and flows unimpeded through the most intensively farmed part of the Penobscot River basin: fields of potatoes, beans, corn, berries, sheep, goats, cows. Six miles from Bangor, the rapids begin, the stream falls over broken mill dams and through coarse ledges where salmon and eels were caught beneath the 150-foot tall cliff known as Lover's Leap. On January 3, 1828, Mr. Timothy Colby, the notable fisherman, while fishing for frost fish [tomcod] in the Kenduskeag, caught "a fine fat salmon weighing five and a quarter pounds. This was the only salmon, probably, ever caught in that way through the ice in winter."

Where sawmills once shipped 250 million board feet of lumber in one year, and hillsides boomed with wealth while cash-fat river drivers brawled with roustabouts in the bar- and brothel-filled streets of Hell's Half-Acre, the waterfront is now filled with oil storage tanks, a railroad switch yard, parking lots, and office buildings. Signs of a city trying to shed its industrial past and reinvent itself.

Where, beyond the city boundaries, forest stretched for millions of acres in an unlit vastness that inspired Thoreau to declare Bangor a "star on the edge of night," now spread the gravel roads of industrial loggers and real estate prospectors.

Where the river once ran rank with sewage equivalent to the waste from millions of people, thousands now paddle each spring in the annual Kenduskeag Stream Canoe Race, and allow themselves to be drenched in water that residents once feared. Where once stood six mill dams, each with a fall of ten to fifteen feet, today the canoe race demands portages and navigating rapids in the final six miles before the Kenduskeag empties into the Penobscot.

Mills moved or went out of business, and their owners abandoned the dams; only two barriers remained when fisheries biologist Richard Cutting visited the Kenduskeag as part of his assessment of salmon potential in the Penobscot watershed.

Cutting faced the wave of urban renewal that had come to downtown Bangor. In addition to tearing down nearly half of the city's historic structures, including Union Station—where the first presidential salmon was put on a train to William Howard Taft—city planners wanted to entomb the last half mile of Kenduskeag Stream in giant pipes beneath a new parking lot. Despite the Penobscot being named "a model river" for Atlantic salmon restoration—the last best chance for the species' survival—engineers drew designs for a hot tunnel of cement, a move that according to Cutting represented the end of any potential restoration of Atlantic salmon in the Kenduskeag tributary.

The Bangor Urban Renewal Authority did not entomb Kenduskeag Stream; instead, they narrowed the stream from 250 feet to 80 feet and added parking along both sides.

The Kenduskeag is a flashy stream that responds dramatically to rain and drought. In spring it runs fast and high from snowmelt, but quickly calms to shallow riffles. In the 1960s, when the water level dropped in summer, the stream was more sewage than water. Dissolved oxygen fizzled well below the salmon's required five parts per million. According to Bangor's health department, conditions were ripe for an intestinal disease epidemic.

Raw sewage from 1,643 households—half a million gallons a day—spilled into one four-hundred-yard reach of the Kenduskeag. A thick green scum collected on sand bars, floated in the eddies. At the time, because the stream flowed through a populated area, the Kenduskeag's sewage problem exemplified the kind of mess the new water pollution control laws were attempting to clean up. Congress continued to increase funding for treatment plants, but construction could never keep pace with urban growth.

Ed Muskie turned his undesirable assignment to the Public Works Committee into a command post for environmental action, and his competitive attitude helped drive President Richard Nixon to emphasize environmental policies. With the Penobscot salmon fishery almost nonexistent for much of the 1960s and early 1970s, Nixon never received a first salmon from Maine. Though he supported environmental protection in general, Nixon didn't want to pay for it in a time of war. He vetoed the Clean Water Act as extreme and needless overspending. Congress overrode Nixon's veto in one day, by overwhelming bipartisan margins in both houses. Ed Muskie stood before his fellow senators and asked, "Can we afford clean water? Can we afford rivers and lakes and streams and oceans which continue to make possible life on this planet? Can we afford life itself? Those questions were never asked as we destroyed the waters of our nation, and they deserve no answers as we finally move to restore and renew them."

Of course, driving Muskie and the rest of Washington was an American public who believed the environment was in crisis, who had already witnessed successful grassroots activism, who had faith in technology, and who refused to watch their natural heritage go extinct.

Not since FDR's New Deal envisioned a more planned and scenic America, and before that the conservation era of Theodore Roosevelt, had government been so concerned with ecological issues. Environmental damage—undeniable, finally, with its stinking brown foam, smog-hued sunsets, and disappearing wildlife—was a new issue for most Americans and one that incurred unprecedented involvement by the federal government. The public's interest in the environment had grown, and they pressured agencies charged with governing natural resources to act.

The October 1972 amendments to the Federal Water Pollution Control Act had ambitious goals: eliminate water pollution and restore American waters to their drinkable, swimmable, fishable origins. Instead of asking how much pollution the Penobscot could accommodate, the new law asked how much pollution could be prevented. The act finally provided the funding and enforcement teeth that proved crucial to cleaning up the Penobscot. To start, the newly created Environmental Protection Agency (EPA) gave Penobscot River cities and towns $14 million to build secondary waste treatment plants.

In 1974, with funding from the EPA, University of Maine doctoral student Charles Rabeni began studying the effects of pulp and paper mill pollution on benthic macroinvertebrates, aquatic larve of insects and other tiny bottom-dwelling animals. The aquatic communities at three locations below

Great Northern's East Millinocket mill and Lincoln Pulp and Paper were deemed to be highly degraded, covered in sewage bacteria, and inhabited only by pollution-tolerant worms, leeches, and midge larvae. The river reach was listed as a "Federal Water Quality Limited Segment."

Great Northern Paper Company spent $36 million to install a sulfite recovery system and construct a wastewater treatment plant. They also ended the log drive. In 1971, the final 10,385,465 feet of pulpwood floated down the West Branch Penobscot River, as river-driving gave way to logging trucks. From then on, the river's flow was regulated solely for hydropower generation.

Other mills followed suit, advertising the money they spent on improvements while continuing to stall and resist the new laws. Augustus Moody, environmental engineer with the Diamond International paper mill in Old Town, expressed the sentiment of the mills when he publicly disagreed with zero discharge goals, believing that "a river's assimilative capacity should be used."

The pulp and paper companies still held much power over Maine state government, and they were able to gain exemptions from discharge permits. Anyone dumping waste directly to the river needed a permit that set standards, limits, and compliance schedules for reducing pollution, but a grandfather clause in Maine's 1967 water pollution law exempted all discharges occurring before 1953, and therefore those of the Penobscot's pulp mills, from having to obtain a license under the new National Pollutant Discharge Elimination System.

The logging companies that supplied the mills, meanwhile, were engaged in an enormous chemical campaign against the spruce budworm, an insect that devoured the spruce trees that had come to dominate Maine's industrial forests. Sporadically from 1954 to 1970, and every year until 1985, airplanes sprayed toxic chemicals like DDT and carbaryl over hundreds of thousands, perhaps millions, of acres of forest.

Early in 1973, Maine's new Department of Environmental Protection began to show signs of abandoning its traditional tolerance of paper company pollution, and started demanding more of the mills. Local, state, and federal actions, combined with an estimated $33 million for water treatment plant construction, reduced pollution loading from 1,000,000 pounds of oxygen-demanding waste per day in 1964 to approximately 200,000 pounds per day in 1977. Stoneflies and mayflies came back to the polluted stretches of river; the sewage bacteria disappeared. Pollution had been reduced by 80 percent, prompting the EPA to declare the Penobscot River "a water quality success story."

As the water began to look and smell better, people came back to the river. Hundreds participated in the Kenduskeag Stream Canoe Race. Across New England, the number of canoe manufacturing companies increased from two to twelve. Sales of Old Town canoes increased. The first inflatable raft floated down the Kennebec River in 1976, and the West Branch the following year.

The Department of Interior was studying the West Branch for potential inclusion in the Wild and Scenic Rivers system; Maine's Allagash River had been added in 1970. A team of agency staff and local representatives, including Maine lawyer and rivers advocate Bill Townsend, members of the Penobscot Paddle & Chowder Society, and paper company employees, had toured the river by plane and canoe in the summer of 1974. The Golden Road was under construction, and clear-cutting of the forests was occurring but remained "well screened" from the river. The team camped on Gero Island in Chesuncook Lake, and noted that the shorelines were littered with driftwood.

The Penobscot River was deemed qualified for the federal Wild and Scenic River program in 1976, placing a moratorium on dam construction. This did not sit well with Great Northern Paper executives, who were contemplating building a new dam on the West Branch. Paul McCann, spokesman for Great Northern Paper, announced the company's opposition. "We've been in business here seventy-five years. The very fact that the river is in its present state so that 90 percent of it qualifies [as Wild and Scenic] is a pretty strong endorsement for the way we've run things."

They saw another layer of regulation, adding up to less profit for 3,700 employees and 20,000 stockholders. "Federal designation could deny the opportunity to further develop the hydroelectric potential of the river—in all probability forever," said McCann. "Once designated, the power potential of the river would be lost to society."

Yet all those hikers, paddlers, and campers had gotten to know the Penobscot wilderness, and cleaner streams in their backyards. They would come to be a force that, combined with the motivation of Atlantic salmon anglers, would come to rival the power of industry.

As papermakers and politicians negotiated pollution controls in offices and meeting rooms, dam owners were busy building fishways. Between 1964 and

1973, with $1.3 million provided by the Anadromous Fish Conservation Act and matched by the dam owners, new fishways were added to each of the dams on the lower Penobscot. Al Meister, by then chief biologist with the Atlantic Sea-Run Salmon Commission, scrutinized every fishway in the river. He fought with the engineers over entrance locations and designs.

The fishway effort was encouraged by 1958 amendments to the Fish and Wildlife Coordination Act, which required federal agencies to consider the environmental effects of proposed projects in making regulatory decisions. The Act provided "that wildlife conservation shall receive equal consideration and be coordinated with other features of water-resource development programs."

Subsequent events strengthened these requirements. A 1965 Supreme Court decision expanded Federal Power Commission jurisdiction beyond federal dams to essentially every hydroelectric project in the country. Legal challenges to a proposal for a hydroelectric plant at Storm King Mountain on the Hudson River led to new environmental law groups, such as the Environmental Defense Fund, Natural Resources Defense Council, and Sierra Club Legal Defense Fund, who demanded government accountability. The 1969 National Environmental Policy Act expanded environmental considerations beyond fish and wildlife.

Fishways are pointless if there are no fish around to use them. So Maine worked with the other New England states and federal agencies to reorganize and expand the hatchery program. In the 1950s and 1960s, declines in Penobscot salmon runs had forced the hatchery to once again use eggs from Canada (the Miramichi River), supplemented with eggs from Machias and Narraguagus salmon. Throughout this period, hatcheries stocked only fry and parr, with poor results. In the late 1960s, stocking had switched largely to smolts. In 1974, the U.S. Fish and Wildlife Service constructed an $8.5 million hatchery at Green Lake to produce smolts for the Penobscot, with funding from the Federal Aid in Sport Fish Restoration Program.

Instead of being landed by fly fishermen and sent to the president of the United States, the first fish to return were netted at Bangor and Veazie then transported to the hatchery and stripped of their eggs.

The young fish were then put back in the river. They quickly left and spent several years in the relatively cleaner ocean. This human intervention acted as a kind of life support, sustaining the Penobscot's salmon run while other aspects of the restoration were completed.

By the mid-1970s, water quality had improved to the point that salmon could make it on their own upriver to spawn. With nine new fishways and cleaner water, for the first time in over one hundred years, 250 miles of Penobscot tributaries were open to spawning salmon.

And they came back: 138 fish in 1970, 337 in 1972. Spawning adults and inch-long fry were seen in the headwaters of the Passadumkeag and Mattawamkeag; in Great Works and Sunkhaze streams; in the Pleasant, as far as Ebemee Lake; in the East Branch. Salmon traveled as far as they could in almost every stream. In 1974, a salmon surprised James Fawcett of Orono while he tossed a few practice casts into the Bangor Salmon Pool. Once again, the first fish of the year was front-page news; anglers caught more than thirty of the 587 salmon that returned to the Penobscot that year. Fishermen cleared the alder-choked road to the Penobscot Salmon Club and repaired the dilapidated, mice-infested clubhouse. Membership increased from half a dozen to more than seventy-five.

Baseball legend Ted Williams came back, too, to fish the Penobscot for the first time in twenty-five years. "The 56-years old Williams, slimmed-down and looking like he could still swing a fast bat, bounced across the parking ground and onto the clubhouse steps. . . . Williams looked at the ledger, saw anglers had successfully taken 52 fish thus far, and verbally became ecstatic."

"What a remarkable accomplishment. Just remarkable. What an exciting thing for this city," Williams said, as he stood on a rock at Ryder's Ledge at low tide and cast a number eight fly into the Bangor pool. The salmon fishermen looked different—jeans and down vests instead of wool pants and ties, graphite and fiberglass rods instead of bamboo—but their spirits were the same as in 1912.

Then, during a spring flood in 1977, the river took out a chunk of the aging Bangor Dam. The gap gradually widened, and tidal influence resumed its natural range to Eddington Bend and the foot of the Veazie Dam, three miles upstream. Rapids at Eddington Bend emerged as water levels dropped. The fishermen moved upriver and spread out among the salmon pools: Pipeline, Dickson, Big Rock, Eddington, Station B, Beach, Guerin, Club House, Wringer, and Gravel Bar.

One of the first to fish the new pools was Claude Westfall. Though most people fished from shore, Westfall continued the tradition of fishing from boats. With his twenty-foot Grumman canoe, he could move around to find the action.

Westfall was a confident and persistent fisherman. He grew up in West Virginia, on the West Fork River, and spent much of his childhood fishing for bass, catfish, anything he could catch. He moved to Maine in 1952 to attend graduate school at the University of Maine. Right away, he heard about the Bangor Salmon Pool and soon learned the rules of the Penobscot Salmon Club. The fishing had already begun to decline, though, and he soon gave up.

Back on the river in the 1970s, he fished the Wringer Pool well into the night. His angling skills, and ability to catch fish when no one else could, were both impressive and annoying to his fellow fishermen.

Membership in the Penobscot Salmon Club swelled as word spread and out-of-state fishermen from Massachusetts, Connecticut, New York, and elsewhere paid dues for their chance to fish for Atlantic salmon in American waters. From May to July, doctors and dentists came down to the river on their lunch hour, eating sandwiches while standing on rocks, casting, and could be back seeing patients in twenty minutes with a fish in hand. More serious anglers had to be in the rotation line by 3:30 a.m. in order to secure a prime fishing pool. In 1978, anglers hooked twenty-four fish in one day, the largest haul in a long time. Salmon rolled, twisted, tossed, jumped, swirled, their tails scissoring the surface at high tide. At times the fishing was chaos; unbelievable except to those who were there to see it.

So many people fished and wanted to fish that the historic Penobscot Salmon Club didn't have room for them all. Tired of waiting in line to fish, Westfall and other local anglers founded the Veazie Salmon Club in 1978 atop an old dump just below the Veazie Dam. The Eddington Salmon Club incorporated a few years later.

Salmon even appeared in the Kenduskeag Stream. Several hundred salmon migrating upriver in July 1978, seeking cooler and deeper water, ventured into the Kenduskeag. Hundreds of people, awed by the sight of something so wild, came downtown to watch. Parents brought their children, waited patiently for the chance to glimpse *Salmo salar*. Old men leaned against the railings of the canal along Kenduskeag Plaza and told stories about Bangor's once-famous salmon pool.

The situation soon soured, however. Salmon in the shallow Kenduskeag canals were too available, too tempting. Game wardens from throughout the state were sent to Bangor to protect the salmon from crowds of people who trying to snag the fish with dishpans, rocks, baseball bats, hooks, and nets. After numerous violations, citations, and court-imposed fines, the state closed the Kenduskeag to salmon fishing and began taking another look at all regulations concerning salmon. But the presence of salmon also made Bangor residents look at their city in a new light, prompting litter cleanups and civic pride.

The Atlantic Sea-Run Salmon Commission ramped up stocking efforts, and the salmon seemed to respond, with returns surpassing the one-thousand mark. Old fly patterns were resurrected and new ones designed. Game wardens had to patrol the riverbanks, where as many as 1,000 people fished in a single day. People had to re-learn the rules of etiquette: fly fishing only. All pools fished on a rotating system. Boat anglers should be careful not to crowd shore anglers.

Anglers who were once again catching fish, or catching fish the way their parents had before them, tapped into collective memory to revive customs of preparing and eating salmon, including the Fourth of July ritual. Many baked their salmon in the oven (some baked it whole inside a brown paper bag) and served it with the traditional egg sauce. Any leftover fish would be made into salmon salad. Others smoked their salmon, or cut it into steaks and grilled it.

Signs of change and renewal could be found up and down the Penobscot.

On Indian Island, the Penobscots had joined the Passamaquoddy Tribe in pursuing restitution. With the help of persistent, young, idealistic lawyers, the tribes discovered they had valid claims to several thousand acres of Maine land—land transferred without Congressional approval as mandated by the Non-Intercourse Act of 1790. The Nixon and Ford administrations reversed the federal government's policy of terminating Native American tribes, instead supporting their pursuit of self-governance and defending Indian trust rights. The Indian Claims Commission Act awarded $800 million to Native Americans from New Mexico and Arizona to Alaska. Tribes gained much more than money; in Washington State they regained salmon fishing rights.

President Ford asked his staff for an evaluation of the Maine case, and this is what they found:

> The Penobscot Nation constitutes (and has constituted since time immemorial) a tribe of Indians, that the Penobscot Nation used and occupied an aboriginal territory which included the entire Penobscot watershed in the present State of Maine, together with a major portion of the St. John watershed in the present State of Maine, and that the Penobscot Nation ceded the vast bulk of these aboriginal lands in treaties with the Commonwealth of Massachusetts in 1796

and 1818, and in a purchase by the State of Maine in 1833, none of which has ever been approved by the United States.

The illegal treaties eventually totaled 12.5 million acres of land, two-thirds of the state, including land held by Great Northern along the West Branch and half a million acres along the East Branch. With the deadline for submitting land claims looming, the tribes pushed their case, all the while stating they had no intention of displacing residents. They wanted to settle outside of the courts, but the state showed little interest, boycotting federally arranged meetings to discuss a settlement.

Maine's Congressional delegation (Muskie, Hathaway, Cohen, and Emery) responded with legislation that would terminate the tribes' claims, joining a general backlash against Indians in Washington. In protest, some 2,800 Indians representing seventy tribes marched to Washington in July 1978, some all the way from Maine, in The Longest Walk.

The Justice Department, represented by President Taft's grandson Peter Taft, encouraged the parties to settle "potentially the most complex litigation ever brought in the federal courts with social costs and economic impacts without precedent." In 1979, the U.S. Court of Appeals and Maine Supreme Court upheld the sovereign status of the tribes.

The tribes, meanwhile, feared that if Ronald Reagan was elected, he would make good on his vow to exterminate their federal claims altogether, which compelled them to settle during the Carter administration, perceived to be more supportive of their cause. Then Muskie got appointed Secretary of State and was replaced by George Mitchell, the only state politician whom the Passamaquoddies had ever endorsed. With time running out, the reluctant state and cautious tribes agreed to a settlement.

The Maine Indian Land Claims Settlement Acts preserved the reservations and all lands owned by the tribes, while extinguishing their claims to the rest of the land in exchange for $81.5 million (paid by the federal government). The funds and federal recognition enabled the tribes to buy land, rebuild homes, start businesses, go to college, establish health clinics, and relearn their native language. But some aspects of the settlement, which viewed the tribes as municipalities rather than sovereign nations, would prove incompatible with tribal efforts to protect the river and its fish.

More than 3,000 salmon returned to the river in 1980. People like Roger D'Errico of the Penobscot Salmon Club started talking about reviving the

Presidential Salmon tradition. Opening Day had moved to May 1 in order to protect kelts, or black salmon, leaving the rivers after spawning the previous fall. In 1981 the peapods spread out across the river, and anglers cast above the breach in the Bangor dam. Butch Wardwell landed a fifteen-pounder on a "thunder-and-lightning" fly of his own creation. The fish, however, turned out to be a kelt. Hatchery-reared, its stomach full of smelt, the recovering fish was on its way back to sea after spending the winter in the river. Tradition held that it was the first "bright" or sea-run fish that was sent to the president, and so Ivan Mallett and his fresh, eight-pound sea-run salmon caught later the same afternoon took the honor.

Mallett, 61, went back to his home in Lincoln, where he ran a forestry business, and put the salmon in the freezer. Though he would have gladly eaten the fish, instead he waited for state officials to make the arrangements for presentation to President Ronald Reagan at the White House. His fellow anglers, waited, too, anxious to see the revival of the tradition.

Few, if any, knew that the real work lay ahead, that the incredible salmon runs they had come to know were already poised for decline, and the greatest threats to their tradition lay not in the past, but in the very near future.

Chapter Eleven

Welcome to Salmon City

Friday, May 8, 1987

Governor Joseph Brennan struggled to hold the fish. Drops of watery blood fell from the wet and thawing fish onto the carpet. Maine Senator William S. Cohen, Ivan Mallett and his wife, Phyllis, and state Fish and Game Commissioner Glenn Manuel tried not to notice as Brennan handed the fish to Vice President George Bush in the Roosevelt Room at the White House. Bush promised to share a bite with President Ronald Reagan, who was recovering from a gunshot wound inflicted by an unsuccessful assassin.

Reagan's defeat of Jimmy Carter in the 1980 election was no surprise, considering the psychological state of voters. Americans were still coming to grips with the loss in Vietnam and the waves of refugee "boat people" fleeing communist oppression. Watergate remained an embarassment. After oil shortages and violence in gas lines, terrorism, inflation, and a nuclear accident at Three Mile Island, Americans yearned for reassurance about their place in the world. They wanted to believe in themselves again. And so they chose to invest their faith in a Hollywood actor turned politician who offered eloquent promises and convincing optimism.

It's unclear if Reagan would have eaten the gift of a fish. Even when healthy, Reagan ate sparingly: usually just soup and crackers, and Jell-O for dessert, with iced tea. On occasion he was known to enjoy a hamburger, veal stew, fresh coconut cake. He liked meatloaf and macaroni and cheese. For formal occasions, Nancy Reagan introduced the White House to nouvelle cuisine: small, colorful, perfect arrangements of julienned vegetables, salmon mousse, Grand Marnier soufflé.

Barring a state dinner or other function on the agenda, the Reagans usually went to the small study next to their bedroom and ate dinner from portable tables while watching the evening's three network television news shows tape-recorded by White House staff, which perhaps is where, if Reagan wasn't lying in a hospital bed, Chef Frank Ruta might have served them a meal of dilled Penobscot salmon.

Taste preferences aside, Reagan would have appreciated the positivity symbolized by the Penobscot salmon. Barely three months in office and already a hero, it's unlikely he was aware that the energy policies he inherited had invoked turmoil across the country in places like the Penobscot Valley, where some of Mallet's fellow salmon anglers were mobilizing. A company from Massachusetts had proposed to rebuild the Bangor Dam, and the anglers, who had just tasted life with a healthy river, were not about to see it ruined for yet another shortsighted energy development.

The City of Bangor built the original dam at Treats Falls in 1875 to create a municipal water supply. In 1890 the city added hydroelectric generating capacity to power downtown streetlights, one of the first municipally owned electric light plants in the country. Nine hundred feet of stone-filled timber crib stretched across the river to the Brewer shore, where a fishway allowed some, but not all, salmon to pass. Salmon also swam through a log sluiceway notched in the middle of the dam. The dam enhanced the hotspot that was the Bangor Salmon Pool by slowing down the fish, and the brick buildings of the waterworks made a picturesque backdrop.

The city continued to draw water from the river until 1959, when it became too polluted to drink. The turbines at the dam churned on, generating about 3 percent of the city's annual consumption. Like the rest of the nation, Maine had become dependent on petroleum. Homes and businesses switched from coal to oil heat, and overall energy use increased with growing transportation networks and expanding industries like pulp and paper and textiles. By 1971, public utilities were selling electricity at ever decreasing rates thanks to cheap abundant oil. The city abandoned the hydroelectric plant.

Within a few short years, events half a world away in the Middle East made them reconsider their decision.

When the United States shipped weapons to Israel, Arab countries responded with an embargo on petroleum exports. Oil prices skyrocketed. The crisis affected nearly every American, as they waited in line at gas stations, and every Maine paper mill straining to power ambitious machinery with oil-fired boilers.

In response, the federal government encouraged conservation and the development of domestic energy resources. In Maine, Governor Kenneth Curtis called energy problems the most fundamental issue facing the state. Proposals, both real and rumored, for new oil refineries and electricity generating plants spread along the Maine coast. Various commissions, committees, task forces, subcommittees, and offices addressed "the energy problem." Their reports present conflicting numbers about energy use and demand. The exact numbers—different sectors' shares of the pie, anticipated energy uses—don't much matter now. But within the reports are the origins of a renewed interest in alternative energy sources, and the hydroelectric dam proposals that would dominate salmon politics on the Penobscot for more than two decades.

Twenty hydroelectric dams operated in the Penobscot watershed. The ideal hydropower locations had all been developed during the rush in the early 1900s, but that didn't keep people from looking for new dam sites. The New England Energy Task Force reported in 1976 that significant untapped hydropower potential could be realized through development of eighteen new large dam sites. The Army Corp of Engineers identified ten sites in Maine for potential hydroelectric development, including Arches and Sourdnahunk on the West Branch and Winn and Basin Mills on the lower Penobscot. Maine's Office of Energy Resources saw great potential for hydroelectric dams on a smaller scale, each river and stream doing work on its way to the sea.

What would all of this mean for Atlantic salmon, which had made such a remarkable comeback?

Existing hydroelectric installations would be expanded where possible and the few remaining undeveloped sites would be harnessed, Bangor Hydro-Electric Company president Robert Haskell told the crowd at a 1975 conference on Atlantic salmon restoration. "There is little indication that the current energy crisis will be resolved and society will continue to require maximum production from existing facilities," said Haskell.

In Bangor, where the dam and waterworks had deteriorated to the point of hazard and liability, the city council rezoned the land as commercial and entertained scattered ideas for reconstruction: a new dam, a hotel complex, professional offices, medical suites.

But a new hydroelectric dam did not become a real possibility until President Carter signed the National Energy Act in November 1978, amid a second oil crisis. Part of the law, the Public Utilities Regulatory Policies Act (PURPA), opened up the electrical grid, requiring utilities to accept and pay fair wholesale rates for power generated by independent producers.

PURPA also established a low-interest loan program to assist in financing feasibility studies. In 1980, the Windfall Profits Tax Act provided an additional tax credit for property used in the production of hydroelectric power at existing dams. Maine passed laws to supplement these acts at the state level and streamlined the permitting process. The path was cleared and investors began to respond.

By 1980, the Federal Energy Regulatory Commission (FERC) faced more than 350 private hydroelectric projects in various stages of study, planning, and construction. Hydromania had descended upon Maine. Would-be developers perused the New England River Basins Commission's list of seventy-eight sites where still-existing or former dams could economically produce significant amounts of electricity. Included on the list was the broken and deteriorating Bangor Dam.

In February 1981, as salmon anglers tied flies for the upcoming season, FERC accepted Massachusetts-based Swift River Company's application to redevelop the Bangor Dam. Some of the older dam proposals were still floating around. Even President Reagan's old employer, General Electric, threw its hat in the ring.

"There are plenty of rivers in the state. Why do they have to pick on the one river where maybe, just maybe, we have a chance of restoring a wild run of salmon?" asked angler, writer, and Penobscot Salmon Club historian Roger D'Errico. A rhetorical question, maybe, but not one that could be ignored.

Salmon had come back to the Penobscot River, and people had taken notice. D'Errico had been fishing for striped bass in the river; he didn't think salmon would ever come back. A Postal Service employee who learned how to fly-fish in a Brooklyn, New York schoolyard, D'Errico caught at least one fish every year since landing his first salmon in the Bangor Salmon Pool.

Hundreds of fly fishermen worked the shores and currents surrounding the Bangor Dam, infusing the local economy with their spending on lodging, food, and equipment. In a town that had been dubbed "Salmon City, USA," fish were an obvious obstacle to would-be dam developers.

The salmon fishermen reacted quickly to Swift River's proposal, forming two groups to oppose the dam: the Maine Council of the Canada-based Atlantic Salmon Federation and Friends of the Penobscot River. The fly-fishermen took their own actions to protect salmon, enforcing rules of etiquette, and limiting the number of salmon they killed.

The Friends' president, angler Richard Ruhlin, explained their opposition:

> The federal and state government have spent millions of dollars to restore the once-mighty salmon runs of the river and have, in fact, used the Penobscot River as a pilot project. Now, through government regulation, they encourage an activity that will directly and adversely affect these previous efforts in the direct area where those efforts were showing their greatest return.

Politically savvy Ruhlin worked at both the local and state level to garner support. He convinced his fellow Brewer City Council members to oppose dam redevelopment under safety and economic concerns; Veazie and Eddington soon followed. The dam was falling apart, water rushing through and underneath the carcass of rotting logs. One *Bangor Daily News* editorial called it a "rickety, century-old jumble of crumbled cement, rocks, and log cribwork." Maine Senate President Joseph Sewall, a Republican from Old Town, called the plan to rebuild the Bangor Dam "insane."

The City of Bangor approved a lease agreement with Swift River. But the city's right to generate power was in question, and required legislation to clarify. Opponents used the opportunity to press their cause, filling a public hearing room at the Augusta Civic Center with 250 people and scaring the city into withdrawing the bill. Then, the Friends submitted their own legislation to ban any new dams on the Penobscot below Veazie. The governor opposed the bill, believing, like generations of leaders before him, that hydropower and salmon could coexist.

"It's asinine and crazy to dam up the best fishing river in North America for the little bit of power they'd get," Joe Floyd told a *Washington Post* reporter. "We're talking about enough power to light up the city's shopping mall and maybe one street. That's it." Floyd taught and coached at John Bapst High School in Bangor, was a member of the Atlantic Sea-Run Salmon Commission, and vice president of the Friends of the Penobscot River.

Before 1980, the state gave priority consideration to hydroelectric development, and no process existed for preventing construction where a dam would do more harm than good. But the concerns of anglers and others who had come to know a cleaner river percolated into state energy policy. The state's 1980 energy plan was the first to mention that dams conflicted with other uses and values of rivers, noting that "progress in water pollution cleanup efforts and fisheries restoration and/or establishment are viewed as threatened by hydropower proposals."

In June 1981, the Office of Energy Resources submitted the Energy Policy for the State of Maine. The policy promoted developing hydropower, "an indigenous, renewable source of energy," as fully as possible—but only where the advantages of a dam would outweigh its adverse impacts. To help in determining where these dam sites might be (and improve the approval process for developers), the policy directed state agencies to create a statewide fisheries plan, identify the river stretches and dam sites where fish passage facilities would be required, and establish a schedule for anadromous fish restoration. The Department of Conservation, meanwhile, was to identify river stretches in the state that provided unique recreational opportunities or natural values and develop a strategy for the protection of these areas. This "Maine Rivers Study" was conducted with assistance of the National Park Service.

Later the same year, the Land and Water Resources Council (a cabinet-level group who coordinated across Maine's natural resource agencies) released their own study on hydropower that supported most of the provisions in the Energy Policy:

> The state should clearly identify river stretches for which anadromous fish restoration is planned and protect river stretches which are recreational and natural resources of statewide significance. At each hydropower site, adequate consideration needs to be given to in-stream flow requirements to protect other resources uses, including fish and wildlife.

The New England River Basins Commission, too, incorporated other values of rivers into its evaluation of "the extent to which New England's oil dependency could be reduced through feasible hydropower expansion." But by the time their work was wrapping up, Reagan had taken office and cut the commission's funding. The commissioners rushed to publish their final report, without formal recommendations. They created maps of competing uses, including anadromous fisheries, to help prospective developers select sites where minimal conflict was likely.

The clash over the Bangor Dam brought these "competing uses" to the forefront. Similar conflicts spread throughout Maine, forcing the state to review a complicated and inadequate hydroelectric development policy and evaluate the rivers. As reporter David Platt wrote, "What began with the Bangor Dam rippled statewide: before they were really done with the issue, lawmakers and the Brennan administration had completely re-vamped Maine's law regulating the development of dams."

All of these efforts culminated in the four laws of the Maine Rivers Policy, the result of "most ambitious rivers-management bill ever passed by a state legislature." The 1983 Maine Rivers Act provided guidance for "striking a balance" and "seeking harmony" among various river uses while helping developers focus their efforts away from "protected" rivers to those where hydropower was unlikely to present insurmountable problems, although decisions over rivers had to have maximum benefit to the public. The Penobscot River from Sandy Point (Stockton Springs) to Veazie Dam; the East Branch from Medway to Grand Lake Matagamon and the tributaries Wassataquoik, Webster, Seboeis, Sawtelle, Shin Brooks; the West Branch from Ambejejus Lake to the western boundary of T-3, R-10; and from Chesuncook to Seboomook were among river reaches designated as meriting special protection.

Utilities scoffed at these new additional restrictions and regulatory burdens, fully aware that falling oil and electricity prices, withdrawal of tax incentives, and higher development costs were already slowing the rate of hydrodevelopment.

Maine's attempts to incorporate "conflicting uses," such as salmon habitat, into energy planning, and the organized protests of the salmon anglers and other concerned citizens, represented a growing and strengthening environmental constituency, a coming of age of a movement just entering its second decade and about to face its first real challenge, the Reagan administration.

When Ronald Reagan was a young boy in Galesburg, Illinois, his favorite place in the house was the attic, where a previous tenant had left behind a huge collection of birds' eggs and mounted butterflies in glass cases. Escaping his alcoholic father, young Ronald spent hours marveling at the rich colors of the eggs and the intricate and fragile wings of the butterflies. It was, he later wrote, the beginning of his career as "the Great Naturalist."

His family moved again, to Dixon, a town on the Rock River. Reagan read everything he could about birds and wildlife of the valley. He swam and fished in the river, and made overnight canoe trips.

As an adult Reagan ascended through Hollywood to the governor's mansion in California, he came to prefer riding horses at his ranch over fishing and canoeing. While Reagan claimed that he never lost his "reverence for the handiwork of God," his environmental policies as president of the United States suggest otherwise.

Those who propelled Reagan into office had great faith in "free" (unfettered and unregulated) enterprise. Surrounded by men who, blind to the injustices that enabled their entitlement, believed they were "self-made," Reagan promoted economic survival of the fittest. His major supporters (including his "kitchen cabinet" of oil men, steel magnates, executives of car dealerships, drugstores, department stores, and their wives) had suffered through a decade of environmental regulations that cost them money and time, and they were tired of it.

Environmental policy should be more rational and attainable and less idealistic and pious. Love Canal was harmless. Toxic contaminants measured at trace levels were "imaginary and foreign," the risks exaggerated to satisfy previous administrations' thirst for government intrusion into business affairs. Progress in improving environmental quality had been only "modest," yet the resulting regulation stifled progress, strangled industry, deprived the public of the benefit of new chemicals, drugs, and pesticides, and cost too much money.

Reagan immediately cut the federal budget for everything but defense. He revoked almost all of Jimmy Carter's executive orders relating to environmental and natural resources policies. And he named likeminded individuals to lead the agencies charged with protecting the nation's environment, wildlife, and fish.

Anne Gorsuch of Colorado, who had no experience in environmental policy, would lead the Environmental Protection Agency (EPA), which had responsibility for the implementation of nine major environmental statutes, including the Clean Water Act, which Reagan wanted to abolish. Three weeks after she took office, Gorsuch eliminated the enforcement office and reduced staff by 11 percent. She helped instill a policy of "planned neglect" of hazardous waste sites and slowed Superfund monies, and she supported shrinking federal subsidies for wastewater treatment and state-level enforcement and planning. The only responsibility the EPA undertook was to clear more chemicals for quick use.

James Watt would head the Department of the Interior and chair the Cabinet Council on Natural Resources. Watt had been working to dismantle environmental regulations in Colorado and throughout the West. He quickly fired, demoted, or transferred employees who did not share his fear of wilderness, a fear he soothed by quickly opening up wilderness areas to oil and gas leasing. Critical habitat for endangered species went undesignated. Reagan proposed to withhold from the U.S. Fish and Wildlife Service and other agencies the annual revenue earmarked for salmon restoration.

John Crowell, vice president and general counsel for Louisiana Pacific, would be assistant secretary for natural resources and environment in the Department of Agriculture. The resulting rates of timber harvesting were among the highest in American history.

James Edwards, governor of South Carolina, home to eight nuclear power plants and nuclear weapons manufacturing, and W. Kenneth Davis, of Bechtel Corporation, the world's largest nuclear power plant construction firm, would lead the Department of Energy after Reagan failed to eliminate the office as he had promised to do during his campaign.

Environmental policy might have seemed foolish to some, but by the 1980s it was popular with citizens and entrenched in Washington politics. Fish, wildlife, clean air, and clean water had bipartisan support in the House and Senate. Like the residents of Salmon City, Americans had witnessed renewal of the air and waterways and a resurgence in fish and wildlife. Americans liked the EPA. Reagan's efforts to cut budgets, reduce staff, favor business, ruin wilderness, and water down regulations were unacceptable. Public support for environmental protection increased, and membership in environmental organizations grew.

This sentiment came to influence the Maine Rivers Act. The act prohibited new dams on the Penobscot River below Veazie, but existing dams could be rebuilt if the project "enhanced or did not diminish" identified natural resources and values. Because the Bangor Dam was a "rebuild," Swift River pressed on with feasibility studies.

Bumper stickers multiplied on cars around the valley: "Save Our Salmon—No Dam."

In newsletter columns, magazine articles, and editorials, fly fishermen spread their rallying cry, "From Veazie to the sea, the Penobscot will run free!"

Another section of the river not protected by the Maine Rivers Act was the West Branch Penobscot River between Chesuncook and Abol lakes, where, in March 1984, in the midst of the Bangor Dam controversy, Great Northern Paper Company proposed a $100 million forty-megawatt dam across Big Ambejackmockamus Falls.

The 1970s oil embargo turned energy into the paper industry's top concern. Great Northern spent $122 million to convert oil burners to burn coal and bark, and $50 million on a new, more efficient No. 11 paper machine at Millinocket. Their hydroelectric system on the West Branch, 80 percent of the total storage in the Penobscot Basin and the largest private hydroelectric system in the world, supplied only about one-quarter of Great Northern's energy demand.

They needed more. A new dam, the company claimed, would reduce its consumption of fuel oil by 480,000 barrels per year. The "Big A" dam would turn the West Branch's last stretch of legendary whitewater into a four-mile reservoir.

Planning for the dam had been underway since 1975. The Department of Conservation, after conducting its own study of the Penobscot, urged Governor Longley not to request placement in the National Wild and Scenic River program, and the moratorium on dam construction expired.

Great Northern owned 10 percent of all the land in Maine (nearly all of the West Branch watershed) and, with 4,200 workers, was one of the state's largest employers. Without the dam, Great Northern asserted that "the present and future productivity of Great Northern's entire operation in the state of Maine" might be endangered. Even though the cheaper power would be consumed by a private company, Great Northern argued that the public would benefit from continued employment.

Just as Reagan officials underestimated public concern for environmental quality, Great Northern Paper perhaps underestimated the commitment of river advocates who would emerge in response to its proposal for Big A—a constituency created in part by improved water quality and in part by Great Northern's own tradition of allowing public access to its more than two million acres of forest and river.

Sea-run Atlantic salmon had long ago disappeared from the West Branch, but anglers from Europe and all across the United States came to the river to fish for landlocked salmon and wild brook trout. Many more came to experience the woods and mountains, to see a moose or bald eagle, to canoe the lakes and rivers. And, since the first raft floated down the West Branch in 1977, whitewater rafting had grown into an industry; some 9,000 people rode the West Branch rapids in 1982.

Outdoor recreation groups like the Appalachian Mountain Club and Eastern River Expeditions joined environmental and fishing groups in the Penobscot Coalition to Save the West Branch. The national organization, American Rivers, which had just succeeded in preventing construction of three hydroelectric dams on California's Tuolumne River, joined the effort.

Both the Maine Rivers Study and the New England River Basins Commission had identified the whitewater rapids and scenery of the West Branch as having aesthetic, scenic, geological, ecological, and recreational value. The commission went so far as to note

> it has become clear during the course of this study [1978–1981] that opposition to the construction of new dams at previously undeveloped sites is likely to be substantial. In a region where there already exist in excess of 10,000 dams, the number of remaining free-flowing river segments is limited. Many of these sites are highly valued for their fisheries, and for their recreational, scenic, and other assets. If these assets were lost as a result of the construction of new impoundments, it would not be possible to compensate for their loss.

In courageous testimony, Department of Environmental Protection biologist Matt Scott told federal officials the Big A reservoir would fail EPA's dissolved oxygen criteria for several months of the year, degrading fish habitat.

Although the membership of Friends of the Penobscot River and the Penobscot Coalition to Save the West Branch had little overlap, together the campaigns got people thinking and talking about the river in ways they hadn't before.

On the heels of public hearings for the Big A dam—the longest in Maine history—came a four-day public hearing on the Bangor Dam in July 1985. Hundreds of people attended, most in opposition.

Swift River Company, especially vice president Christian Herter and counsel Angus King, remained steadfast in their belief that the dam was compatible with the passage and catching of fish. They would release more water during the spring migration, build new fishways, and stock additional hatchery fish. They believed in compromise. The city believed in compromise.

The salmon fishermen did not.

They started fundraising in anticipation of the court battles that lay ahead. The threat of electricity rate increases enabled the clubs to form alliances with the Bangor Chamber of Commerce, the cities of Bangor and Brewer, and professional organizations. They got a boost when, in March 1986, Great Northern abandoned their pursuit of the Big A dam. Even though the company had a permit in hand, they decided not to build the dam, citing a drain on financial resources, foreign competition, and problems with state regulatory agencies. Great Northern was unhappy with several conditions of the permit. They did not want to involve the state in energy decisions. They would not guarantee future employment in the Katahdin region; Great Northern president Robert Bartlett had already announced a phase-out of 1,365 jobs, and he refused to commit to the modernization plans previously used to support Big A.

And so the Bangor Dam became the major test of the Maine Rivers Policy's new dam-licensing law. The state Board of Environmental Protection denied Swift River's permit request because a rebuilt dam would "diminish" the significant anadromous fish resource value of the Penobscot River, thus violating the Maine Rivers Act. Swift River and the city asked the board to reconsider. They would not, so Swift River took the case to court.

While many of Reagan's appointees had brief reigns, they had a lasting impact on environmental policy, and it took a while for the backlash to catch up. Congress took up the consideration of fish, wildlife, and recreation years after Maine had codified the Maine Rivers Act. Since 1965, when federal license requirements were expanded to nearly all hydroelectric projects in the United States, not just federal projects on navigable waters, the Federal Energy Regulatory Commission (FERC) had gone about their business, with little or no consideration of the environmental impact of their decisions.

In passing the Electric Consumers Protection Act of 1986, Congress noted that FERC's performance had been "less than satisfactory." In several egregious instances, Congress noted, "licenses were issued and concrete poured before fundamental concerns for in-stream flows, fish and wildlife protection, recreation, and energy conservation were even addressed," as required by the Fish and Wildlife Coordination Act, the Endangered Species Act, and National Environmental Policy Act.

The new law directed FERC to give environmental values "equal consideration" with power development, and heed recommendations received by federal and state fish and wildlife agencies. These new requirements, combined with Reagan's elimination of federal funding, ended the era of big dam projects. The act also created new public participation requirements, allowing citizens, agencies, and organizations to comment on license renewal applications, a standard that would come to play a critical role in future deliberations over Maine rivers and Atlantic salmon.

With licenses for 245 hydroelectric facilities on rivers throughout the country scheduled to expire by 1993, the eyes of the nation were on Maine, and the Penobscot River salmon that were at the heart of the question, what do we want our rivers to be?

In May 1987, President Reagan was swapping fish stories with eighty-seven-year-old salmon angler George Fletcher in the Oval Office. The ceremony had moved away from the White House ever since Brennan's sloppy presentation in 1981, but returned to a place of high honor in 1987 in order to give Reagan some good publicity amid the Iran-Contra hearings. Senator George Mitchell was called away from the hearings to attend the ceremony.

"He took it in the spirit of how we in Maine give it," said Fletcher of the salmon, wrapped in paper to prevent leaks. "When we give anybody anything, we mean it. If they don't want it, that's something else."

Reagan liked to talk about the return of Atlantic salmon, the California condor, the peregrine falcon, and "many other magnificent creatures." His executive branch touted local cuisine as a national treasure, such as serving Columbia River salmon at a state dinner for Russian President Mikhail Gorbachev and his wife, Raisa.

But Reagan and his administration failed to recognize that comebacks, or even the persistence of a species, were made possible by the very kind of government action he detested: massive funding for wastewater treatment, air pollution controls, scientific monitoring, compliance enforcement. Instead, he repeated the chorus that protecting and preserving the environment conflicted with an American quality of life.

Salmon anglers on the Penobscot, many of whom were local, working-class folks, were not among those who benefited from Reagan's policies. In 1988, first fish catcher Neil Donnelly refused to turn over his salmon. "President Reagan just cut my income in half," said Donnelly, a logger. "He didn't need my salmon, too. I've had two good meals off it and I'm going to have another tomorrow."

Instead, Charles Caron of Brewer got to send his eight-pound Atlantic salmon to the White House, escorted by Governor John R. McKernan Jr., along with members of Maine's congressional delegation. Throughout the summer, dedicated salmon anglers waited for the Bangor Dam lawsuit to make its way through the courts.

Legal deliberations came to revolve around the meaning of the phrase "no diminishment." To diminish something is to lessen its size, importance, authority, reputation, or value. In November, the Maine Supreme Court upheld the Board of Environmental Protection's decision against the dam—the proposed mitigation would not compensate for the loss of wild salmon. A new dam would diminish the salmon resource, and the Penobscot River. The Bangor Dam would not be rebuilt.

Any celebrations were short-lived.

Chapter Twelve

Alamoosook

Below Bangor and the famous salmon pool, the river deepens, the banks steepen into ironbound cliffs, hung with cedar and pine, and several short turns give the river a gorge-like scenery that once earned it the nickname "The Rhine of Maine." Houses are far apart and set back from the river atop the cliffs and eroding sandy bluffs. The first tongue of saltwater wedges north at Winterport.

When the tide and wind align in a certain way, waves stack up in the narrows. Silvery waves of spartina grass shimmer from pockets of marsh between the cliffs; wide, shallow coves of mudflat glisten where tributaries join the main channel: Souadabscook Stream, Sedgeunkedunk Stream, Cove Brook, Marsh River.

The river, now a brackish estuary, widens and splits around Verona Island. The eastern channel flows past the Bucksport paper mill site and waterfront, into a confluence with the Eastern or Orland River, a wide tidal tributary that flows out of Alamoosook Lake. High ground surrounds the lake. Extensive red ochre burial sites around the shores of the lake testify to the area's importance to the ancient Wabanaki, for whom Alamoosook provided passage north to inland ponds and flowages, and south to the ocean via Blue Hill.

On the eastern side of the lake, a small stream flows over coarse sand and bedrock ledge, the outflow of Craig Pond, half a mile above. When Charles Atkins found the pond and brook, fed by cold underground springs, he knew the crystal-clear water was ideal for raising salmon. Atkins leased land at the mouth of the brook from Elisha Carr and David Dodge in 1871. An old shingle mill provided a structure to begin hatchery work.

Charles Atkins bought hundreds of salmon every summer from the Verona and Stockton Springs weir fishermen, who collected salmon with flannel-lined dip nets. For transporting fish to the hatchery, Atkins designed special salmon boats or "cars," half-submerged dories with square holes in the bottom and sides to allow water to flow, and nets across the top kept fish from leaping. Once a day, the "collecting steamer" *Agnes* made a tour of the weir district, towing cars full of salmon five miles to a series of dams and locks at Orland. Hatchery staff then rowed the sluggish cars upstream to Alamoosook. At the north end of the lake, Atkins built an impoundment to raise the fish until spawning season, when he mixed their eggs and milt by hand, let the adults go (after tagging them, of course), and transferred the eggs to the hatchery at the outlet of Craig Brook. The eggs incubated for five months, in stacked trays also designed by Atkins. He had turned the brook into a series of pools to raise the young fish after hatching; fish were raised in troughs lined up on the lawn.

Atkins's techniques were based on those in Europe, but he made great advances and documented much of his work.

A century later, the Craig Brook National Fish Hatchery underwent renovations as the salmon restoration effort expanded. Equipment became more sophisticated, allowing technicians to haul and release millions of fry throughout the major tributaries of the Penobscot. Construction of a second hatchery at Green Lake had quadrupled the number of smolts being produced to hundreds of thousands. When the returning salmon count on the Penobscot reached 1,000 for the first time in decades, state salmon biologists and managers and the hatchery crew joined in celebration at Alamoosook Lake, with *Bangor Daily News* outdoors writer Bud Leavitt as master of ceremonies.

Launching the modern salmon restoration program during the years when the Penobscot was so polluted was possible because hatchery-raised salmon smolts released into the river each spring immediately left to go to the ocean, bypassing all the issues associated with degraded and inaccessible freshwater nursery habitat.

When smolts hit salt water, they accelerate, swimming up to twenty-five miles per hour, schooling near the surface and traveling night and day. They travel through the estuary in a few days, drifting with the current and then swimming out with the tides. Their progress is neither steady nor unidirectional: they reverse and head upstream periodically with the tide, or wait it out. Smolts take advantage of the relatively high flows of spring, riding the currents to pass rapidly along both sides of Islesboro and into the bay.

Their time in the estuary is brief but crucial. More than half of them won't survive. Some will be eaten by the double-crested cormorants that wait at the mouths of streams and dive from atop old logging booms and pilings. Others will be eaten by otters, eagles, osprey, gulls, herons, seals, porpoise, cod, pollock, striped bass. Some won't successfully transition to salt, because they left too early or too late or somewhere along the route were exposed to pollution.

Smolts are especially vulnerable to low pH, which interferes with the enzyme that pumps salt across the gills. Acid rain has devastated salmon in some areas, such as Nova Scotia, but it's less of a widespread issue in Maine, where scientists have had a harder time separating acid rain effects from the natural acidity of the landscape. Maine rivers, including the Penobscot and especially those to the east, are naturally acidic due to the bedrock and extensive peat bogs and evergreen forest. Decades of acidic precipitation may have depleted the ability of some watersheds to buffer against periodic acid conditions. In a few of these locations, under certain conditions, low pH can affect salmon smolts.

Those smolts that do survive, their endocrine hormones heightened by the smolting process, continue to imprint, their olfactory nerves recording organic matter in the current, bookmarking scents along the migration corridor so they can find their way back. For a long time, their marine phase was considered "a black box."

In their early experiments, Isaak Walton and Charles Atkins showed that salmon left their spawning grounds and returned, but where they went in between remained a matter of conjecture. In 1892, A.N. Cheney—"the well-known angling expert and writer on fish-cultural matters"—noted in *Forest and Stream*, "Where the salmon go when they descend the fresh water rivers and enter the sea is as yet a matter of speculation. There is a certain mystery about the habits and movements of the sea salmon after it has left the fresh water rivers in which it spawns, and gone down to the sea, that never has been satisfactorily explained."

Knowledge of salmon at sea came sporadically, and then only from places where fishing occurred. Cod and pollock fishermen reported catching salmon in trawls and purse seines. U.S. Fish Commissioner Hugh Smith reported the occasional appearance of salmon in mackerel and herring weirs off Matinicus and Cranberry islands. "The salmon are usually taken during the time when the fish are running in the rivers, but occasionally one has been caught in midwinter. On June 19, 1896, a Gloucester fishing vessel brought into Rockland a ten-pound salmon that had been caught on a cod trawl twenty miles southeast of Matinicus . . . several salmon have been taken on hooks off Frenchman Bay."

Thirty years later, not much more had been learned. William Kendall was repeating the same refrain: "Where the salmon lives while in the sea is a subject of much discussion, and a question which has not been fully answered." Kendall noted more reports of salmon caught off Nova Scotia and Cape Cod.

Twentieth-century salmon biologists kept trying to follow salmon in the ocean. When their research efforts resumed after World War II, they began more tagging studies. In October 1956, a tagged kelt released in Scotland's Blackwater River was recaptured by coastal fishermen 1,700 miles away near Sukkertoppen (Maniitsoq), West Greenland.

In 1962, Al Meister started tagging Maine salmon, using a modified Carlin tag. Developed in Sweden and named for its creator, Börje Carlin, the tag had a tiny metal plate engraved with a unique number. Meister attached tags using a black polyethylene carpet thread, instead of the usual forked wire, and needles special-ordered from a factory in Belgium. He recruited nimble-fingered women to produce the tags at the Craig Brook hatchery, and designed a special cradle to hold the fish while tagging.

The following summer, tagged salmon showed up in mackerel and herring weirs along the south shore of the Bay of Fundy. The next year, fishermen in Nova Scotia and Newfoundland found more fish, and one even showed up in Greenland. Tagged salmon (most had spent one winter at sea) showed up off the coast of Labrador and Newfoundland, and in the gillnets of salmon fishermen on the western coast of Greenland, in cod and mackerel traps, and other saltwater weirs, as well as back in the Penobscot, providing the first concrete evidence of their distant migration. Al Meister and Richard Cutting tagged tens of thousands of American salmon, part of an international effort demonstrating that American and European salmon populations converged to spend their winters in the food-rich waters near the Arctic Circle.

Greenlanders had seen salmon before. The earliest documented captures of Atlantic salmon off Greenland date from the early 1600s. Atlantic salmon sometimes spawned in the River Kapisillit, which is the Greenlandic name for salmon. Limited fisheries for salmon took place during the fall in the early twentieth century, but most Greenlanders were fishing for cod. Cod fishermen also reported catching salmon in the coastal waters of West Greenland in the 1930s, but a targeted fishery did not begin until the 1950s.

Various explanations exist for why commercial fishing started at this time.

Farley Mowat, in *Sea of Slaughter*, wrote that during the late 1950s, the U.S. Navy began sending submarines north under the Arctic ice, and one of them made a surprising discovery: "gleaming hordes" of Atlantic salmon beneath the Baffin Bay ice. "First to make capital of the discovery were the Danes, closely followed by the Norwegians. A fleet of deep-sea seiners and driftnetters was soon butchering salmon for the first time on their wintering ground."

Most accounts begin the targeted fishery in the 1950s. Greenland, a colony of Denmark, had relied on the cod fishery for sustenance and income, but cod catches had been declining. Denmark invested large sums of money in the country, part of a vast program of modernization to increase Greenlanders' standard of living. The Royal Greenland Trading Company provided funding to the locals so they could buy boats and nets to see what they could get to eat. They set nets within fjords and in and around the rocky shore. They found salmon, which they consumed fresh or salted, as freezing facilities had yet to be built.

Away from the coast, in the high seas of the North Atlantic, vessels from the Faroe Islands and Norway began fishing with drifting gill nets three to twenty miles long. Local fishermen in Greenland and Newfoundland quickly adopted the new, nylon monofilament nets, expanding their own fishing offshore. Construction of modern fish-processing plants provided employment for a growing population that was finding seal and cod stocks declining. Atlantic salmon brought high market prices, and the equipment and infrastructure to catch, preserve, and ship fish had improved. Catches rose quickly through the 1960s as more vessels joined the frenzy, from an estimated twenty thousand in 1960 to nearly one million in 1971.

The West Greenland fishery caught a mix of North American and European salmon. Half a million salmon were being netted before they could return home to spawn, and fisheries biologists and managers on both sides of the Atlantic had to figure out which fish were being caught. A catch of 100 salmon could represent one fish from 100 different rivers, or 100 fish from a single river.

Many feared that the sudden multinational, complex fishery would cause a collapse in salmon returns and ultimately of the species, which was already showing signs of population decline. Government scientists, angler groups, and other salmon advocates pushed for international cooperation. The International Commission for the Northwest Atlantic Fisheries agreed to recommend that fishing by foreign vessels be phased out, and established quotas for the Greenland vessels. The high seas fishery decreased after 1971, but Atlantic salmon were still being caught throughout their range.

In Canada, where the economic impact of recreational fishing far outweighed that of the commercial fisheries, the government decided to favor the recreational fishery and phase out commercial fishing by "buying out" more than 5,200 commercial licenses between 1972 and 1976, with more over the next decades, at a cost of $70 million Canadian.

When the new Magnuson Stevens Act extended U.S. jurisdiction two hundred miles from the coast, putting an end to the International Convention for the Northwest Atlantic Fisheries, non-governmental organizations including Restoration of Atlantic Salmon in America proposed a new international organization, and pressured the U.S. government to take action to protect salmon. Their call was echoed at a symposium in 1978, when the Atlantic Salmon Trust and Atlantic Salmon Federation introduced an international treaty agreement to ban ocean fishing for salmon. The draft treaty became the Convention for the Conservation of Salmon in the North Atlantic Ocean, and at the urging of President Reagan, the U.S. Senate ratified the Convention in 1983.

The resulting North Atlantic Salmon Conservation Organization (NASCO) focused at first on debating and developing regulatory measures, and establishing quotas.

NASCO meetings were politically charged and tense, even poisonous, although resolution came with time. The International Council for the Exploration of the Seas, which provided scientific advice to NASCO, consistently said that too many salmon were being killed, but for years no one listened. Very powerful political forces in Europe were still making money off the salmon and wanted their commercial fishery. Compromise sometimes happened in the middle of the night, sometimes aided by whiskey. Canada's commercial fishing moratorium provided an example of a country making sacrifices for their salmon resource, helping other countries reduce and even end harvests. Some nations sold their quotas to conservation organizations.

Maine salmon biologists, thrust into international politics, traveled to Greenland to meet their salmon on the other end of the migration. Al Meister recalled one trip, when a gracious host, thinking she was giving her guests a real treat, served salmon to the unhappy biologists. "When you've done nothing but handle salmon your whole life, you don't want to eat them," said Meister, whose database of millions of salmon tag returns was instrumental in international negotiations to end commercial fishing at sea.

The high seas fishery raised issues of sustenance and resource rights, issues that were being tested at home, on the Penobscot River.

The formal and informal proceedings surrounding the Bangor Dam proposal offered the Penobscot Indian Nation the first chance to assert their newly reaffirmed sustenance fishing rights.

In 1987 the tribe announced plans to gillnet up to twenty salmon from their reservation lands on the Penobscot River for ceremonial purposes in connection with the next year's Indian Days festival. Like an echo from the Taft era, sport fishermen opposed the plan, calling gillnetting a method from "the dark underbelly of fishing."

When Native Americans first began ceding their lands to the United States, they did so with the assurance that they could continue fishing, hunting, and other lifeways. The Penobscots retained their aboriginal fishing rights granted by early treaties, rights that were confirmed by Maine law in 1979. But the tribe and the state, with different interpretations of the Settlement Acts, continued to disagree. The state claimed that land flooded by dams was no longer part of the reservation, or that the Penobscots could not legally take fish from the river.

Sea-run and marine fish and shellfish made up as much as 35 percent of the traditional Penobscot diet, along with berries, roots, shoots, nuts, tubers, and fungi. The Penobscots may have been prevented from engaging in a fully traditional lifestyle at the time, but that did not mean the status quo was in any way acceptable. They wanted to pursue their rights to practices that included the use of natural resources; not the lifeways of people with semi-suburban or hybrid lifestyles and grocery-store diets.

In the summer of 1988, the Penobscots netted two salmon from the river. The fish were served as part of the tribe's annual community celebration. But, given declines in salmon numbers, they decided not to take any more salmon out of the river. They also had come to realize that the fish might be too polluted to eat.

In 1982, a Penobscot wildlife manager and a game warden launched their boat at South Lincoln for a routine patrol. They noticed a sheen on the river and followed the leak to the nearby Lincoln Pulp and Paper mill. Fuel oil was pouring into the river from a malfunctioning oil-water separator. No one was around, so they went up to the mill to tell someone about the leak. They later reported the incident to the Department of Environmental Protection.

John Banks, a forest ecologist who had just helped assemble the Penobscot Nation's Department of Natural Resources, was shocked. How often does this kind of thing happen? he wondered. What are the impacts on the health of the river, of tribal members? Stories had been circulating about people getting sick, illnesses related to environment conditions. The incident raised a lot of questions about how water quality was being managed. Under Banks's leadership, the Penobscots developed a water resources program and built a state-of-the-science laboratory.

Clean, healthy water is inseparable from tribal culture. Preventing pollution, an act of tribal self-preservation and sacred stewardship of their home lands and waters, cannot be entrusted to others. The Penobscots began monitoring acid rain, mercury, and toxic contaminants in animals and plants with ecological and cultural importance: fish, muskrat, snapping turtle, freshwater mussels, fiddlehead ferns. They sampled dioxins and furans in sediment at the bottom of the river.

Dioxin was one of their most serious concerns. Lincoln Pulp and Paper released 2, 3, 7, 8 tetrachloro-dibenzo-p-dioxin (TCDD), a by-product from bleaching wood pulp with chlorine and chlorine dioxide, into the river thirty miles upstream from Indian Island. Dioxin is the most potent known carcinogen and impairs human ability to reproduce. In 1987, the state issued its first fish consumption advisory, advising men to limit their intake of fish from the Penobscot River, and cautioning women and children not to eat *any* fish from the lower river due to dioxin and mercury contamination.

The tribe's monitoring documented that the Lincoln mill regularly violated its discharge permit, including some massive spills of dioxin. All along the river, data collected by the Penobscot Nation—thousands of environmental samples from hundreds of locations in the watershed every year—led to the discovery of expired permits, falsified documents, broken wastewater pipes, and leaking fuel tanks. They documented elevated levels of mercury in parts of the West Branch, as well as increasing levels of phosphorus coming from the Millinocket mills that were fueling extensive, toxic algae blooms, including one that consumed enough oxygen to cause a fish kill.

Their data also helped to upgrade classification and ensure higher standards for more than four hundred miles of the Penobscot River. The Penobscots monitored the river intensively—much more so than the state could afford on its own. With less staff turnover, the Penobscot Nation water resources program had extensive knowledge of the river, and they worked with state agency staff, collaborating on research and sharing data.

But state leaders resisted any tribal influence over water quality, at first claiming the Penobscots would be too lax in their regulation, and then that they would be too stringent. Behind the state's view were the polluters who had long prospered in the shadows of the status quo. While Maine's pulp and paper industry had been waning since the 1970s, the industry was still a large employer in an economically depressed state. In the 1990s, mills were still making pulp and paper at Millinocket, East Millinocket, Lincoln, Great Works, Brewer, and Bucksport, with wastewater discharged at each location.

As writer Julia Bayly explained, "The last thing the state wanted to see was a sovereign nation living on a river. . . . What could be more terrifying to those whose profits depend on discharges than people for whom clean water in that same river was a top priority?"

By the end of the 1980s, the high seas fishing fleets could not meet their allowable quotas—Atlantic salmon had become too scarce. Salmonid stocks were declining worldwide. On the Columbia, the negative impacts of some twenty-five dams had accumulated to reduce migratory fish by 75 percent. More than 200 Pacific salmon and steelhead stocks faced a risk of extinction. On the Penobscot, the salmon population was almost entirely dependent on hatchery stocking.

The dramatic and widespread decline in salmon populations, and subsequent increase in market prices for fresh salmon, prompted many people to start thinking about other ways to maintain salmon as food. Hoping to reduce pressure on wild fish, the Atlantic Salmon Federation promoted the newly emerging salmon farming industry.

Atlantic salmon aquaculture began in the fjords of Norway and the lochs of Scotland in the 1960s, building on hatchery methods developed for wild stocks. The industry started in Maine in 1970; in 1983 and 1985 Craig Brook National Fish Hatchery provided smolts to a private company in order to "jumpstart" the Maine aquaculture industry. The first Maine-grown fish reached the market in 1987. The industry expanded rapidly. Farm-raised salmon from Maine, Canada, Norway, and Chile flooded markets, surpassing wild fish in percentage of world supply. With supply greater than demand, prices for fresh salmon dropped, making the fish accessible and popular. Salmon consumption more than doubled.

While aquaculture did succeed in relieving pressure on wild fish, the practice introduced a whole new set of complications for salmon restoration. In the early years of the aquaculture industry, Atlantic salmon grown in Maine and the Canadian Maritimes were a mix of strains from the Penobscot, St. John, and various Norwegian strains from forty-one different rivers, giving them thirty to fifty percent European genes. Unintended releases of farmed salmon have occurred in every region where the fish are cultured. In Maine, cultured fish were documented in the St. Croix, Dennys, Narraguagus, and Union Rivers. Farmed fish, more aggressive, faster growing, and larger than wild fish of the same age, competed with wild salmon for food and habitat. Eventually, better equipment and containment procedures, restricting strains to those of North American origin, and local reductions in production contributed to a massive decrease in the numbers of escaped farm salmon in rivers.

Aquaculture also distanced consumers from the source of their food. In Maine, the salmon anglers and their families were the only ones who knew the taste of wild Atlantic salmon. But as salmon numbers declined and dam proposals threatened their sport, anglers started talking about catch-and-release, even doing away with killing a fish for the president. When Don Shields caught a fish in 1992, he released it, but felt that he still deserved to have a salmon. So he went to the store and bought a piece of farmed Atlantic salmon. The move away from the river as a source of food was nearly complete.

Chapter Thirteen

Paper Salmon

Monday, May 1, 1989

Bill Ellison, 41, landed the first Atlantic salmon of the season shortly after 6 a.m. Monday in the Wringer Pool on a firebug fly. Most of Ellison's fellow anglers knew him as Salmon Billy, and described him as a "roundabout." Mostly homeless and often drunk, he lived in a tent along the river's edge near Veazie. He spent most of his time fishing, or getting others to fish, hurrying them along their rotation through the pools. "Stark raving mad," blowing his whistle, and yelling, "C'mon! You've got to get out there fishing! Be more productive!" One story tells of a tipsy Ellison swimming from one side of the river to the other to get to where the salmon were biting, fly rod clenched between his teeth.

Inevitably, Salmon Billy caught the presidential salmon. There was his picture on the front page of the *Bangor Daily News*, eyes rolling behind his glasses, big grin on his face, wild beard and long hair sticking out from under his baseball cap, clutching his fish, all five pounds of it. Some eyebrows must have raised. *This* person was going to present *that* puny salmon to the president?

Indeed he was, but Salmon Billy wouldn't go to Washington unless his (slightly estranged) son, Yancy, could come along. At 4:30 a.m. Salmon

Commission biologist Ed Baum went to the Bangor airport and met the Ellisons: clean and neat, wearing new suits that Don Shields had bought them for the occasion. On the flight down, Salmon Billy may or may not have had a few drinks.

The presentation of the president's salmon was brief. At some point, maybe while they were waiting at the White House or in a lobby afterward, somewhere in proximity to a piano, Salmon Billy revealed that he was a trained concert pianist.

Baum bought lunch, and a couple of beers for Billy. He got home at 1:30 a.m. the next day, and vowed that he would never, ever go on another presidential salmon trip. He would much rather be out on the river, with the fish.

President Bush also preferred being on the water to ceremonial formalities. A few weeks after receiving his first salmon as president, Bush arrived in Maine to spend a long weekend at Walker's Point, taking advantage of the unseasonably warm weather to speed *Fidelity* out on the Atlantic for an afternoon of fishing.

Within months of the defeat of the Bangor Dam proposal, Bangor Hydro-Electric Company officially unveiled their long-in-the-works plans for a new 1,640-foot-long, eighteen-foot-tall concrete dam at Ayers Falls in Orono.

After dividing around Orono and Marsh (*wasahpskek*, "slippery ledge") islands, the Penobscot River again becomes a single channel, a series of bony ledges and rocks create a wide, shallow staircase of whitewater, and a good place for taking fish.

This same fall of water lured James Walker, who purchased a new mill and in 1853 built a dam eight hundred feet long and fifteen feet tall to power an expanded sawmill complex, in its time the largest sawmill in the world under one roof. Walker used the "basin" between a small island and the western riverbank to hold six million feet of logs waiting to be cut by two gangs of saws and sixteen single saws, and machined into clapboards and shingles. When the water was low, the dam shunted the whole flow of the Penobscot into the basin under the mills, water wheels groaning and saws screeching.

In the 1860s, when dams in Bangor and Veazie were either low enough for salmon to leap over, or had fishways or other gaps, the best salmon fishing was often below the Basin Mills Dam.

By 1900, sawmills shared the basin with a box factory and the Orono Pulp and Paper plant, constructed on the island in 1889. The "Basin Mills" burned down in 1910, and high flows washed out the dam in 1936. The river eroded the visible parts of the dam; the pulp and paper mill reopened in 1938 for a few years and was then converted to a shoddy (wool scrap fiber) mill.

That the new $125-million-dam would be an additional barrier to migrating salmon was obvious. The dam would back up the Penobscot for about three-and-half miles, all the way to the next dams at Orono and Great Works.

Bangor Hydro's plans included doubling the power produced at Veazie and relicensing other dams to serve the utility's 110,000 customers. The company claimed the dam would favor electricity consumers and create hundreds of jobs during the construction phase. They believed the benefits far outweighed the cost of losing a wild salmon run.

Bangor Hydro proposed what they thought was a simple and acceptable solution to replacing the fifty adult salmon they estimated would be lost as a result of flooded habitat. The fishways designed by federal biologists were optimistic and too expensive; instead, Bangor Hydro planned to trap and truck salmon around the dams, and build a fish hatchery closer to the river. In fact, claimed Bangor Hydro, the dam would actually improve habitat for fish like shad and smallmouth bass by making the channel deeper.

According to the U.S. Fish and Wildlife Service, whose authority to influence dam licenses was strengthened by the 1986 Electric Consumers Protection Act, if Basin Mills Dam were built, the probability of restoring salmon to the Penobscot by 2043 dropped from more than 70 percent to below 40 percent.

To the Penobscot River salmon clubs, the Penobscot Nation, and others, Basin Mills Dam would be the final nail in the coffin. Turning more of the river into a reservoir risked lowering dissolved oxygen concentrations below what salmon could tolerate. Fish populations would never recover with yet another blockade in their path, no matter how much money Bangor Hydro was willing to spend on chutes and ladders.

Those who were not exhausted from fighting construction of the Bangor Dam and the Big A Dam refocused their efforts on the new threat upstream.

Don Shields, Bob Milardo, and John Diamond met over Bob Wrengzyk's garage and formed the Penobscot Coalition. Then, on December 2, 1988, Don Shields gathered some members of the Maine Council of the Atlantic Salmon Federation in the Eddington Salmon Club to discuss how to raise awareness of the threat posed by Basin Mills, and to prepare for the conflict to come. "The public is unaware that there is opposition to Basin Mills," Milardo told the group. "They tend to accept the dam as inevitable." Bill

Townsend had just become president of the Maine Council, and had been thinking about the future direction and focus of the organization. When Basin Mills was announced, he knew their course was set.

Based on the experience opposing the Bangor Dam, the Penobscot Coalition knew they would need money to support their efforts. *Bangor Daily News* outdoors columnist Bud Leavitt paid a visit to the Veazie Salmon Club, where President Claude Westfall and other members were having coffee. Leavitt, according to Westfall, was a catalyst. He had the idea for a fundraising banquet, and he volunteered himself and Westfall as co-chairs. The $100-a-plate "Stop Basin Mills" event in September 1989 featured baseball legend (and avid salmon angler and Leavitt friend) Ted Williams and *New York Times* reporter Nelson Bryant. With this and other events, $10 and $100 at a time, they raised $125,000 over a period of four to five years.

But the fishermen had a harder time convincing people that the dam was a threat, because it wasn't going to be located on top of the Bangor Salmon Pool, but a few miles upriver sandwiched between two existing dams at Veazie and Great Works. They feared that without the support of major statewide organizations like the Natural Resources Council of Maine and Maine Audubon (who were busy trying to remove the Edwards Dam on the Kennebec River), the regulatory community would not view the issue as important. They recruited others to their cause: the Sierra Club, Conservation Law Foundation, and American Rivers, who listed the Penobscot River as the third most endangered river in the nation in 1989.

Eventually, the Sportsman's Alliance of Maine and the Natural Resources Council of Maine, each with about 9,000 members, went on record opposing Basin Mills. Trout Unlimited signed on.

To the Penobscot Nation, which was just getting an environmental monitoring program up and running, Basin Mills would interfere with their rights to harvest sea-run fish in the river, rights they hoped to be able to exercise again someday. They expressed concern that by causing more delay in migration, the dam would affect the quality of salmon, since the more time fish spend in freshwater, the more their flesh deteriorates. The Penobscots wanted Bangor Hydro to pay compensation for loss of traditional fishing resources.

"Our position is simple," said Paul Bisulca, chair of the Penobscot Nation's hydro review committee, in a public hearing,

> We want returned to us sufficient runs of salmon, shad, and alewife to be able to harvest them in our waters. We are not anti-development, but in this case the potential harm to our fisheries outweighs the benefit of this project to us and

those people of Maine who inhabit the mid and upper areas of the Penobscot Valley who should be able to take for consumption those fish that we want restored.

The U.S. Department of Interior consistently supported the Penobscots' aboriginal rights to fish in the Penobscot River. The State of Maine consistently denied the Penobscots rights to fish in the river.

The salmon anglers were wary of tribal control of the fishery; the tribe resented the fact that anglers could catch and eat fish but none reached their reservation. Eventually, both sides realized they had something to gain by working together, although the tribe never joined the Penobscot Coalition, preferring instead to represent itself.

Those in support of the dam—the Maine Chamber of Commerce and Industry, Associated General Contractors of Maine, and a few paddling guides and paddlemakers, who believed they would benefit from more flatwater recreation—organized, too, forming the Friends of Basin Mills.

The stage was set for a classic drama, with salmon front and center.

Bangor Hydro primed the public relations machine. Folders of fact sheets and glossy brochures reassured that "clean water and fish passage technology make it possible for fish to thrive," while acknowledging the federal stocking program that "brought back enough Atlantic salmon to create a modest sport fishery" and boasting that the Penobscot offers "the best smallmouth bass fishery east of the Mississippi." The information packets featured such illustrations as a man fly-fishing from a canoe, and a "Native American Artifact"—a wooden salmon spear. Energy consumption is on the rise, advised a tri-fold brochure on thick paper with blue and silver ink. The "low profile" dam would offer "better fisheries" and "more recreation." Scattered technical details intended to impress or overwhelm: dam specifications, power supply estimates, details on fish numbers and passage plans.

Bangor Hydro liked to show graphics of how many dams already existed on the river. They liked to talk about how the river had been used to provide energy for "almost two centuries." They claimed that compared to current handling methods, their plan for fish passage would accelerate the Atlantic salmon restoration. They did not talk about the results of studies at their dam in West Enfield showing 86 percent of downstream migrating smolts were drawn through the turbines, with unknown numbers killed.

Bangor Hydro's proposal was based on its own projections of increased demand. But at the time, there was an excess of electricity in the region.

Bangor Hydro believed the existing supply glut would disappear, and if it didn't, they would abandon their plans.

The campaign worked. Friends of Basin Mills swelled to a nonprofit organization with more than three thousand members who believed that being in control of rivers was a good thing. Its members didn't like that a small group of people "who claimed to be fishermen" would deny others the right to safe, clean, low-cost electricity.

The fishermen produced their own materials. *Penobscot River Renaissance*, a $50 hardcover, limited-edition book published in 1992, featured interviews with salmon anglers and river advocates:

> Sometimes I long for the good old days, when all we had to deal with was the fish. We'd just go out and do our field work. Today it's meetings, meetings, meetings and "paper" salmon! —Ed Baum

> The Atlantic salmon and the rivers go hand in hand; I can't see how anyone could love the Atlantic salmon and not be in love with the rivers, or love the rivers and not the salmon, since the salmon are synonymous with a good river. —Dick Ruhlin

> We're restoring the salmon for the sake of the fact that they should be allowed to survive. . . . If there's some angling effort or enjoyment that some can get from that, that's fine, but that's not why we're doing it. —Jerry Marancik

> This river has already given to the maximum, with all the hydro on it. There should not be any more dams on the Penobscot. What the power companies are hoping is that eventually the restoration agencies in the public sector will give up and go away and they'll have the river to themselves. —Bill Townsend

> It is discouraging after all the work, money and effort that has been put into the restoration program to find ourselves now faced with Bangor Hydro's plan to build another dam. . . . Once you have something taking place on the river and man interferes, by building a new dam or fishway, it's just not the same and it will never be the same. The entire ecosystem is affected. . . . Unfortunately, biological diversity, a loss of plant and animal species, has become one of our most pressing problems. The future may not appear to be very bright at times, but we must continue to oppose those projects that threaten our quality of life. —Claude Westfall

One year after Claude Westfall delivered the first salmon to President Bush at Walker's Point and returned to protest Basin Mills, his son Scott caught the first salmon of 1993. New rules restricted anglers to keeping one fish per season. Scott put his fish in the freezer and waited for the politicians to organize the presentation, which would be made to Vice President Al Gore instead of President Bill Clinton. Three months later, he was still waiting. Tired of the run-around, Scott declared, "I'm going to eat the fish." He defrosted the salmon, cooked and ate it, sharing some with his dog.

Salmon had become, as Ed Baum said, a thing on paper rather than an animal in the river. A procession of salmon task forces, boards, authorities, working groups, and committees marked the end of the twentieth century. Accompanied by an entourage, protocol, and an expensive journey, the delivery of the president's salmon had become arduous. The spirit of the gift was crushed by the politics of its presentation, just when the salmon most needed the attention.

That fall, the state Board of Environmental Protection voted in favor of the Basin Mills Dam, and a group of citizens petitioned the federal government to put the Atlantic salmon on the Endangered Species List.

The U.S. Fish and Wildlife Service and National Marine Fisheries Service considered listing Atlantic salmon as threatened or endangered in 1991, but a petition from RESTORE the North Woods, Biodiversity Legal Foundation, and Jeffery Elliot forced the agencies to review the status of *Salmo salar*. In 1995, they proposed a threatened listing for Atlantic salmon in the Sheepscot, Ducktrap (a Penobscot tributary), Narraguagus, Machias, East Machias, Pleasant, and Dennys Rivers. They needed more time to study the Penobscot salmon.

Just as they feared outsider control of the state's water in the 1910s and 1920s, state leaders negotiated with the federal agencies to try their own way.

In their proposed rule, the agencies encouraged the state to take the lead and create its own conservation plan. The state reorganized the Atlantic Salmon Commission, developed a long-term river-specific fry stocking plan, completed habitat assessments of critical spawning and nursery areas, and formed watershed councils for each listed river. The aquaculture industry established codes to limit salmon from escaping from farms. In 1995, salmon anglers were limited to catch-and-release fishing only. The state provided anglers with a farmed-raised fish to replace the wild salmon they could no longer keep and eat.

Meanwhile, the Atlantic Salmon Federation, the Penobscot River Coalition, and other conservation interests, assisted by the Sierra Club Legal Defense Fund and financially backed by Trout Unlimited, Orvis, and American Rivers, appealed the Board of Environmental Protection decision in favor of the Basin Mill Dams, but both the Maine Superior and Supreme Courts upheld the state's approval of the dam.

Bangor Hydro still needed approval from the Federal Energy Regulatory Commission (FERC), however, which could only be obtained after a full environmental impact statement. While this was being prepared, Basin Mills again landed the Penobscot River on American Rivers' list of the nation's ten most endangered rivers. The Penobscot made the list eight years in a row, more than any other waterway. Salmon anglers and the Penobscot Nation found support emerging from federal agencies that had poured millions of dollars—and hours—into restoring the Penobscot over the previous decades, including the agencies reviewing the status of the species, as well as the Environmental Protection Agency, Department of Interior, even the Army Corps of Engineers. The changing dynamics of the deregulated electricity industry, sustained by the 1992 Energy Policy Act and subsequent state laws, forced utilities to separate power generation from distribution. Bangor Hydro would have to sell its existing dams, but the company pushed on with their plans for Basin Mills.

In 1998, after a skeptical final Environmental Impact Statement, FERC denied Bangor Hydro a license for Basin Mills. In addition to costing three times more than power from other alternative sources, the new dam would adversely affect fish migration, accelerate the decline of Atlantic salmon, and render impossible a "self-sustaining" run accessible to subsistence and recreational consumers.

> Federal and state agencies and Indian tribes responsible for managing the natural resources of the Penobscot River Basin have recognized the negative impacts of past developments on the river, including hydropower dams, and have invested substantial time, money and effort in implementing plans to protect the natural resources affected. We conclude that the proposed Basin Mills development would frustrate the goals of and be inconsistent with plans to protect Atlantic salmon. . . . We conclude that the construction, operation, and maintenance of the Basin Mills development, even with the proposed mitigation and enhancement, is not best adapted to the comprehensive development of the Penobscot River for beneficial public uses. The application for license is denied.

And with that, the dam-building days on the Penobscot were over.

Salmon, however, remained in the headlines. The state's efforts to prevent endangerment (and Endangered Species Act listing) of Atlantic salmon proved inadequate. In 1999, 1,452 salmon returned to New England rivers, 968 to the Penobscot, a decline of 60 percent from 1979. In late 1999, the federal agencies proposed listing Atlantic salmon as endangered in the Dennys, East Machias, Machias, Pleasant, and Narraguagus (the "Down East Rivers"), as well as Cove Brook and Ducktrap River (tributaries of the Penobscot), and Sheepscot River.

The state legislature passed a resolution opposing the listing. The Eddington and Veazie salmon clubs opposed the listing. All angling, already limited to catch-and-release, ended with the listing proposal. At business breakfasts and civic meetings across northern Maine, the listing was debated. State and local leaders urged hundreds of people to attend public hearings and oppose the listing. They feared regulatory uncertainty, rampant unemployment, loss of forestry, aquaculture, and agriculture. At the Machias hearing, Senator Olympia Snowe called the listing "a Draconian proposal" that would result in economic decapitation.

Many argued that after centuries of hatchery production and mingling with escaped farm fish, Maine's Atlantic salmon were nothing but "mongrels," synthetic artifacts unworthy of protection.

Edward French, editor of *The Quoddy Tides* newspaper, tried to counter the acrimony:

> If salmon are not surviving, then our activities are part of the cause for their precipitous decline in population. You cannot simply blame the seals. Clearly this is a call to continue efforts to clean up watersheds. And if salmon are faring so poorly, what does that mean for other species—including ourselves? In the Pacific Northwest, salmon are held in a hallowed position, but not so in Maine . . . aside from anglers, there are few who will lobby for better management for the salmon populations. . . . Perhaps we can show that extinction is not acceptable. . . . Without greater emphasis on maintaining and restoring the environment, we will encounter more difficulty supporting our natural-resource based economy . . . the debate over listing need not be acrimonious. It should focus on how all involved can work together to restore habitat—not only for the salmon but for all of us.

Governor Angus King, who worked for the Swift River Company when the Bangor Dam was rejected, found himself once again in a snarl over salmon.

Joined and supported by the Maine State Chamber of Commerce, Wild Blueberry Commission of Maine, Maine Forest Products Council, Bangor Hydro-Electric Company, Penobscot Hydro LLC, Maine Pulp & Paper Association, Maine Aquaculture Association, Atlantic Salmon of Maine, LLC, Stolt Sea Farm, Inc., Jasper Wyman & Sons, and Cherryfield Foods, Inc. and FPL Energy, LLC, King sued to prevent the listing.

Fueled by news reports that ESA listing of Pacific salmon was having "a profound effect on people's lives," the business interests dreaded a "heavy-handed regulatory burden" and feared national environmental groups would use the law to force changes in their businesses.

Chapter Fourteen

Wassumkeag

Below Orland and Verona, the estuary becomes Penobscot Bay. A large island drops from the western shore, a forested mound of glacial till with bedrock outcrops at the 200-foot summit. Tiny streams tumble over wooded cliffs onto sand and gravel beaches that gave the island its name, *Wassumkeag*, landmark of bright shining sand to canoe travelers.

Later, people knew the island as Brigadier's Island, for General Sir Samuel Waldo, and then Sears Island, for the Sears family, who used the island as a summer estate. Cows, pigs, and sheep grazed the meadows, and salmon were netted from the western shore. U.S. Fish Commissioner Hugh Smith reported that a Sears Island fisherman caught 408 salmon in 1896, the largest catch of the year. The salmon fishery extended throughout the inner bay, to Rockport, Islesboro, and Brooksville.

Salmo salar is "king," according to the writer Ted Williams, "because it has a monopoly on the qualities by which people judge fish." Large, sleek, and silver, the Atlantic salmon swims great distances; returns after years at sea; and leaps seemingly insurmountable falls to get to its home waters. When caught on a fly, a salmon resists, making for challenging sport. When cooked and eaten, salmon provides toothsome protein, fat, vitamins, and minerals.

Historical abundance of Atlantic salmon in the Penobscot never approached the run sizes of the multiple species of Pacific salmon, which in part explains why Northeastern cultures are not as salmon-centric. Instead, salmon were one of a "suite" of sea-run or "diadromous" species, each with its own run timing and successive season.

At any given time of year, ten other species move through the Penobscot on their way to and from the ocean: fish large and small, short and long, in shades of blue, brassy green, gray, silver, black. In spring, out-migrating salmon smolts are passed by alewives, blueback herring, shad, and smelt moving in the opposite direction. Striped bass, sea lamprey, and Atlantic and shortnose sturgeon arrive in summer. In fall, juvenile alewife and blueback herring leave the river, and mature yellow eels journey to the Sargasso Sea. Rainbow smelt and tomcod, or "frostfish," swim beneath the ice in winter. Many of these species also supported sustenance and commercial fisheries.

Wabanaki place names, historical records, and maps attest to the distribution and abundance of migratory fish.

Blackman Stream, *Alahkecimehsihtek*, place where there is little fishing with casting nets. Mattimiscontis Stream, *Matamehskahtis*, the little alewife location. *Passagassawakeag,* the name of the river that enters Penobscot Bay at Belfast, place for spearing sturgeon by torchlight.

European settlers fished from the eastern shoreline of French (Treat-Webster) Island along "Shad Rips." Salmon Stream, a tributary far up on the East Branch, points to a former spawning ground, as does Shad Pond, on the West Branch near Millinocket.

All of these fish are still present in the Penobscot River, but at a tiny fraction—around 1 percent—of their historic numbers. They disappeared due to the same changes that affected salmon—dams, fishing, pollution—and their loss likely contributed to the declines in Atlantic salmon. Or, as fish biologists say, the river's natural diversity of co-evolved fish species may have offered salmon "demographic security."

For example, smolt out-migration (tens of thousands of fish) coincided with the in-migration of alewives (millions of fish), which are similar in appearance and higher in calories, and therefore more likely to be eaten, helping to "buffer" salmon smolts against predation. In the same way, adult shad entered the river at the same time as adult salmon, and may have provided an alternate food source for seals and other large predators. Atlantic salmon kelts regained strength by feeding on rainbow smelt.

Other species act as habitat engineers. In their late spring nest-building and spawning activity, sea lamprey clean and loosen gravel streambeds, priming the area for the autumn spawning of Atlantic salmon.

The millions of fish that once swam up the Penobscot brought ocean nutrients into upland watersheds, and imported protein throughout the food web, from seals to bald eagles. Through their excreted wastes, spawning activity, and death, diadromous fish contributed to the health and robustness of the Penobscot ecosystem. In West Coast watersheds, such links are well-established. Who knows how much sea-run fish were responsible for the towering height of King Pine?

The same principle holds in reverse. The flux of fish out of the Penobscot River supported the marine ecosystem, including the nearshore Gulf of Maine.

"You name it, we caught it," local fisherman Avery Bowden told the Maine Folklife Center in a 1978 interview. Bowden once trapped salmon, and occasionally sturgeon, alewife, blueback herring, skate, tomcod, squid, hake, haddock, mackerel, dogfish, eel, and shiners in his weir at the head of Penobscot Bay. Even more species showed up in the nets and weirs of other salmon fishermen: Atlantic herring, shad, menhaden, striped bass, cunner, tautog, sculpin, lumpfish, cod, pollock, hake, sturgeon, lamprey, smelt, mummichog, and flounder.

Racks of split and salted cod dried in the sun at Bucksport, once a major port for cod and other groundfish. Cod spawned in the channel west of Islesboro and off Castine; haddock spawned fifty miles inland off Sears Island.

Nineteenth-century officials noted the connections between river and sea fish.

In his 1873 report to the U.S. Commission of Fish and Fisheries, Spencer Baird documented his conversations with fishermen about the impact of declining forage fish on groundfish populations in eastern Maine:

> That period [of large catches] was before the multiplication of mill-dams, cutting off the ascent of the alewives, shad, and salmon, especially the former . . . the reduction in the cod and other fisheries, so as to become practically a failure, is due to the decrease off our coast in the quantity primarily of alewives and, secondarily, of shad and salmon, more than to any other cause.

By 1880, he assumed that the reduction in the number of the anadromous fish was "the chief agency in the decrease of the ocean-shore fisheries."

The co-existence of salmon and other migratory fish in the rivers, towering forests, and an ocean full of cod is not coincidence, but a kind of co-dependence.

In the 1990s, restoration efforts began to consider a larger range of species, and a broader view of river ecosystems. On the Kennebec River, the Edwards Dam removal targeted not salmon but striped bass, rainbow smelt, alewives, and sturgeon. Plans for the Penobscot River also came to encompass the diadromous suite, and to focus on the whole ecosystem extending out into the Gulf of Maine, in part because focusing only on salmon wasn't working.

In the time between the dramas over Bangor Dam and Basin Mills, salmon populations in the Penobscot and elsewhere drastically declined, raising questions about the methods and purpose of restoration, and the purity of salmon stocks.

The Board of Environmental Protection favored the Basin Mills dam because the existing salmon run was already dependent on human intervention. With or without the new dam, they believed restoration of a "completely self-sustaining salmon run" unlikely. Some stocking would always occur; the run would never be completely wild.

Many people accepted this. Some anglers who cared more about fishing than the fish thought a "put and take" fishery just fine.

Others refused to accept defeat. *Bangor Daily News* writer and angler Tom Hennessy wrote, "It's impossible to moderate, ease, mollify, or justify the loss of anything natural. Anyone who thinks otherwise is either totally naive or in it for the money." The U.S. Fish and Wildlife Service and Atlantic Salmon Commission wanted a self-sustaining run of salmon, which the Basin Mills project would prevent.

This had long been the goal. When the success of the salmon hatchery at Craig Brook became evident in the nineteenth century, U.S. Fish Commissioner Baird emphasized that the real objective was "to establish a continued run to be kept up by naturally spawned fish, a result which should be continually aimed at."

Were Penobscot River salmon truly "wild" fish, or mere hatchery products requiring constant human assistance? When the U.S. government listed a "distinct population segment" of Atlantic salmon in eight Maine rivers, including two Penobscot tributaries, as endangered in December 2000, the state dissented, claiming genetic continuity from aboriginal to contemporary salmon stocks could not be "conclusively demonstrated." Through state and federal hatchery programs, Maine argued, humans had been manipulating salmon genetics for over a century, and the idea that salmon returning to Maine rivers were distinct lacked scientific support.

A population is considered a distinct population segment if it represents an "evolutionarily significant unit" of the species. Their strong homing tendency means that salmon readily diverge into genetically different and reproductively isolated "populations" or "stocks" with local adaptations, including the timing of spawning runs, appearance, and other features. For example, Maine Atlantic salmon have a higher proportion of adults returning after only two years at sea (more than 80 percent in Maine and fewer than 50 percent in Canada and Europe). Maine Atlantic salmon grow faster and smolt earlier (after only two years), whereas Canadian salmon may take two to five or more years to reach smolt size. Within the Penobscot watershed, West Branch salmon likely differed from East Branch salmon and those that spawned in the Piscataquis, Passadumkeag, etc.

In determining whether or not to list any or all of Maine salmon under the Endangered Species Act, biologists from the U.S. Fish and Wildlife Service and the National Marine Fisheries Service considered the population's persistence, the history of fish stocking, the geographic segregation from other salmon populations, and genetic differences. They concluded that the influence of hatchery fish wasn't strong enough to completely or substantially hybridize with the remnant populations of wild salmon.

But the state and other businesses, believing the listing would damage the economy, disagreed. They argued that the distinct population segment designation was not based on the best data, was overly broad and unduly vague, and unconstitutional. Any unique genetic legacy in the listed rivers had been overcome by stocking. Federal officials invented the "wild" label so they could apply their own agenda.

Unable to fish, the salmon club members felt angry, abandoned, disillusioned, and increasingly apathetic. Most disagreed with the listing.

Maine's congressional delegation, meanwhile, asked the National Research Council to study the matter.

The resulting interim report on genetics concluded that

> North American Atlantic salmon are clearly distinct genetically from European salmon. In addition, despite the extensive additions of nonnative hatchery and aquaculture genotypes to Maine's rivers, the evidence is surprisingly strong that the wild salmon in Maine descend from a common ancestor geneti-

cally distinct from Canadian salmon. Furthermore, there is considerable genetic divergence among populations in the eight Maine rivers where wild salmon are found.

In 2003, U.S. District Court held that the determination of endangered status was sound—and supported by the facts on record.

Remnants of wildness persist. Glades of knee-deep ferns conceal old stone walls and foundations. Deer trails vanish into alder thickets. Huge silver birches grace the meadows. Tides flood the salt marsh; waves flatten the stones of the beach. Harbor porpoise and seals arc and dive offshore. Gulls and osprey keen overhead.

At 941 acres, Sears Island is one of the largest uninhabited islands along the entire East Coast of the United States. The five miles of accessible shoreline are rare in Maine, where much of the waterfront is privately owned. Sears Island is one of few places to skip a rock across the water, find a piece of driftwood.

Yet the island is also nothing special. Its location is the primary reason for its notoriety. Bangor & Aroostook Railroad bought the island, and settlers left a few decades later in 1934. Over the next few decades, the island reverted to a wilder state; until 1969, when developers proposed the first of many intensive uses for the island: an aluminum smelter, an oil refinery, a nuclear power plant, a coal-fired power plant, a bulk cargo and container port for shipping potatoes, wood chips, and paper from the northern hinterlands.

The latter effort advanced far enough to construct a long causeway along what had been a sand bar exposed at low tide, permanently connecting Sears Island to the mainland, and to pave a road up the spine of the island down to a rock jetty now overgrown with weeds.

Citizen protest, backed by legal action from the Sierra Club, twice blocked construction of the port. As with Basin Mills, ever-stricter regulatory requirements from the U.S. Fish and Wildlife Service, the National Marine Fisheries Service, and the EPA revealed that the project was incompatible with fish, wildlife, and functional ecosystems. Angus King, governor of Maine, conceded another victory to environmentalists in February 1996, after spending more than fifteen years and around $20 million on the failed Sears Island port.

The environmental impact statements associated with the proposed port and other developments repeatedly cited Sears Island as being "not unique," "not unusual," "characteristic," and "common." Even scholars called the island "unremarkable." Bisected with a road, its wetlands ruined, clam flats poisoned, electric wires strung through the forest to a radio tower that blinks in the night, Sears Island is no wilderness. Across the cove, the oil storage tanks and shipping terminal of Mack Point hum and clatter. But remnants of wildness persist.

In 2003, rumors of a proposed liquefied natural gas terminal on Sears Island drew attention to the island's perpetual vulnerability. Governor John Baldacci, not wanting to repeat his predecessor's experience with Sears Island, charged a group of municipal, state, and nonprofit interests with finding a compromise.

It was a season for compromise.

One year after FERC's denial of Basin Mills, Pennsylvania Power and Light of Maine, LLC (PPL) purchased Bangor Hydro-Electric Company's dams in Maine, a result of electric industry deregulation, and became the owner of all the hydroelectric projects on the lower Penobscot River.

The company did not want to deal with continued "rancor" over each and every dam. They wanted to resolve the contested dam relicensing and fish passage issues in a more holistic way. Over the next few months, they met with the Penobscot Nation, the agencies, and other conservation interests. Everyone came to the table, but they couldn't agree on a solution.

Negotiations became fragile. John Banks knew he needed to do something to remind everyone why they were there, to help them see beyond their own needs and interests. He came to what could have been the last meeting and asked for a little bit of time at the beginning of the agenda for a small ceremony. He said a short prayer, and then took an eagle feather wrapped in red cloth. He went around the room, tapped each person on the shoulder with the eagle feather and said, "This is not about you. This is about the river."

The tone of negotiation changed, and the parties soon reached an agreement. PPL and FERC would delay relicensing to give the conservation interests (organized as the Penobscot River Restoration Trust) five years to raise $25 million to purchase the Great Works, Veazie, and Howland dams. Once purchased, the trust would raise another $25 million to remove the two lower dams and build a bypass channel around Howland Dam on the Piscataquis. In return, PPL would be able to make modifications to their other dams to mitigate loss of hydropower, "rebalancing" the river's energy generating capacity.

Given all the issues with the habitat, biologists knew they couldn't just keep throwing fish at the problem. Restoring access would certainly help the Atlantic salmon. Another solution was to maintain genetic diversity.

In 2003, new genetics data showed that Penobscot salmon (and those in the Androscoggin and Kennebec rivers) were similar to the endangered salmon in the eight rivers listed in 2000. The National Marine Fisheries Service and the U.S. Fish and Wildlife Service ultimately expanded the endangered listing of Atlantic salmon to cover the entire Gulf of Maine distinct population segment, including all of the Penobscot.

The listings forced a major change in the philosophy and approach to hatcheries. Rather than making more fish for aquaculture or sport fishing, or even restoring a self-sustaining run, the hatcheries focused solely on preventing extinction. Protocols changed from those of a production facility to a live gene bank with a goal of preservation: collecting parr from each river, raising them to adults (in separate, locked rooms), tracking individual fish from egg to adult, analyzing mating pairs with computer programs designed to maximize genetic variability, stocking fish back into their home river.

Hatcheries staved off extinction (97 percent of adult returns in the Penobscot are from stocked smolts) but at the expense of diversity. With each generation that was smaller than the one before, accumulated lifetimes of variation disappeared, never to be replaced. Diversity provides a reservoir of options for future adaptation. Maine salmon migrate the farthest distance to West Greenland, and a high percentage stay at sea for two winters. Fish from each river and tributary may have special characteristics that could help the species survive. Fish that smolt earlier survive better in warmer temperatures. Fish from long rivers like the Penobscot might develop and behave in a different way compared to fish in short rivers. Others take different migration pathways—in the past, some Maine salmon may have used the Bay of Fundy and the Gulf of St. Lawrence. A lot of river-specific fitness has been lost already. Who knows what kind of unique traits were carried by the salmon of the West Branch? The question now is whether enough variability remains in the endangered stocks to create resilience. Adding urgency to this question is the rapidly changing ocean.

After they pass Sears Island and leave Penobscot Bay, salmon post-smolts move in a south-south-easterly direction across the Gulf of Maine. At the mouth of Penobscot Bay, river discharge and offshore winds typically push the westerly flowing Eastern Maine Coastal Current toward the central gulf, helping the young salmon along in their migration.

Within a month, post-smolts reach the coast of Nova Scotia; by June or July they are nearing Newfoundland. They swim at the surface, eating krill and other energy-rich microscopic crustaceans, growing fast to avoid predation by sharks, skate, halibut, whiting, and cod. As they grow they eat more fish: juvenile herring, sand lance, capelin. In late summer and early fall, some will go as far as West Greenland. As winter approaches, they head back to the southern Labrador Sea to await the spring migration home. The following spring, some males (an average of 20 percent) return as "grilse" or "one sea winter" salmon; most stay at sea for another year.

The anadromous life cycle has costs. Transitioning between fresh and salt water is stressful. Swimming long distances uses energy and time, and there are more predators in the estuary and ocean. An anadromous life cycle also makes salmon challenging to study. Even when highly abundant, salmon constitute only a minor component of the marine ecosystem: needles in a watery haystack.

Work by the National Atlantic Salmon Conservation Organization (NASCO) and non-governmental organizations ended the commercial high seas fisheries. Since 2000, only West Greenland has harvested salmon, for subsistence only. While initially the shutdowns of commercial fisheries bolstered adult returns, Atlantic salmon kept declining on both sides of the Atlantic, especially fish from the southern edge of the range. The estimated abundance of North American Atlantic salmon declined by 85 percent, from 800,000 fish in the mid-1970s to fewer than 110,000 fish in 1998; northern populations did better than southern fish.

The poor survival of salmon at sea coincided with a climate change-driven "ecosystem reorganization." In the early 1990s, the North Atlantic Ocean experienced a regime shift. Melting Arctic ice sent a pulse of cold, low-salinity water into the North Atlantic, diluting salinity and leading to early stratification and changes in currents. In the Gulf of Maine and Labrador Sea, phytoplankton prospered in the warm, isolated surface layer, fueling small, faster-growing zooplankton with less fat and nutrition than the Calanus and other species preferred by capelin and other small fish eaten by salmon.

Capelin disappeared. Cod stocks crashed off Greenland. Other fish, including potential predators of post-smolts, shifted their distributions in response to increased temperatures as the Gulf of Maine warmed at incredible speed. Whiting, red hake, dogfish, monkfish, sea raven, and rosefish all increased in abundance.

Other, large-scale climate factors influence the salmon's marine environment. Atmospheric pressure systems and their related precipitation, wind, air, and sea surface temperatures have patterns on decadal and multi-decadal scales. Any and all of these changes—more or new predators, lack of food, changing currents and wind patterns, could explain the crash in salmon populations.

Climate changes affect salmon in freshwater, too.

Smolts use day length (photoperiod) to initiate the physiological changes that prepare them for life at sea, and they tend to leave when temperatures reach 50 degrees. As freshwater rivers have warmed, salmon smolts are leaving earlier and over a shorter period of time. But leaving too early risks arriving into an ocean without food, or with unfamiliar predators.

Food web changes are believed to be a more important factor influencing salmon declines than warmer ocean temperatures alone. Salmon can tolerate warmer water than originally thought, and they can adjust their feeding behavior. And salmon, given their past experience with advancing and shrinking ice sheets, have some ability to adapt to changing climate conditions. Atlantic salmon have adjusted to changing habitat over the course of their existence as a species, including the landlocked salmon's adaptation to not migrating. As the water warms, Atlantic salmon will be one of the very few diadromous species able to tolerate the challenging climate of Quebec and Labrador—their glacial affinity will become a benefit.

The Atlantic salmon is *septentrional,* which means "of the north" and refers to the seven stars of Ursa Major, the Great Bear, or Big Dipper that dominate the night sky over the Northern Hemisphere. If the Arctic ice continues to melt, salmon may expand into the newly exposed territory. Atlantic salmon were able to survive after being introduced to the far off barren Kerguelen Islands near Antarctica. Penobscot River salmon represent the southernmost genes, which might hold important ways to survive at the southern edge of a warming salmon world. Dealing with such change is likely encoded in their genes.

Of course the chance exists that *Salmo salar* could be wiped out by abrupt climate change or even more drastic reorganization of the ocean ecosystem than what occurred in the early 1990s, but the Atlantic salmon has proven its resilience. But for the species to survive in the Penobscot River, salmon will need access to more habitat.

More salmon with access to more habitat increases the likelihood of straying. Although strays probably have lower reproductive success than fish that are returning to their native streams, they do provide some potential for new genetic combinations, and perhaps more importantly, they allow for re-colonization of streams if a local population has disappeared. As salmon start to use more of the habitat, they can adapt to a greater variety of environmental conditions; when they reproduce, they expand genetic diversity. The key is that they need access to their home waters, and they have to find conditions suitable once they arrive.

According to the Endangered Species Act, "recovery" does not mean restoring salmon to their historical abundance and distribution (tens of thousands of fish spread throughout the Penobscot watershed), but instead merely getting the population high enough to be taken off the Endangered Species List (thousands of fish in some of the watershed). Because there was enough habitat in the rest of the watershed to meet recovery goals, the agencies did not include the West Branch Penobscot River as critical habitat.

As the Penobscot River Restoration Project neared realization, partners looked more carefully at the entire watershed and the different migratory fish species. Sturgeon, striped bass, and other "lower river" species would regain nearly all of their habitat as a result of the dam removals. Other fish, like alewives and shad, would readily use newly available habitat, and alewives in particular could be "jump started" by stocking reproductive adults from other watersheds. But salmon are different. Salmon (and eels) are the hardest to restore because they need nearly the entire watershed.

More than a few dam removals are needed to restore the continuum of stream networks and reconnect freshwater and ocean ecosystems. The factors that contributed to the decline of Maine's salmon—private ownership of the watershed by timber, pulp and paper companies—may be the very things that end up saving Atlantic salmon. Extensive forest cover throughout the watershed has protected water quality. More than 100 dams in the Penobscot watershed continue to block fish passage, but compared to other rivers of its size, the Penobscot has relatively few major dams or other developments. The Penobscot continues to be a "model river" for Atlantic salmon restoration.

The tradition of giving the first salmon of the year to the president of the United States kept alive a vital connection between people and the surroundings that sustain us. The salmon anglers maintained a cultural memory of Atlantic salmon as food and wild animal. Many salmon anglers have passed away, and no young anglers are able to take their place in the Bangor Salmon Pool. Claude Westfall and most of the other salmon fishermen doubt they'll ever fish again.

Members of the Penobscot Nation do not exercise their rights to harvest Atlantic salmon. To prevent possible harm from mercury, dioxins, and PCBs due to eating freshwater fish, the tribe advises children under the age of eight, and women who are nursing, pregnant, or may become pregnant to eat no fish from the Penobscot watershed.

"The culture suffers when we can not practice the old ways and pass them on to future generations, Tribal elder Butch Phillips wrote. "The culture suffers when we cannot learn the lessons of the ways that have sustained our people since time immemorial, and the People are weakened when the young no longer seek the wisdom of their elders. The culture suffers when practicing the traditions can cause harm to the health of the People."

At least two generations have passed since people have lived in proximity to large runs of Atlantic salmon. And yet Americans are eating more salmon than ever before. Most consumers don't distinguish among the multiple farmed and imported fish marketed as "Atlantic salmon." In *American Catch*, Paul Greenberg argued that without valuing rivers and oceans as a food source, we can't understand why they need our defense. If we can't eat wild Atlantic salmon, and indeed have come to prefer farmed salmon, why do we need clean, unimpeded rivers and all the other environmental preconditions that make wild salmon possible?

Even if salmon never return to the Penobscot because of climate or other unforeseeable changes, making the river healthy enough for salmon will help us, too. Imagine an environment that is hospitable to salmon: Tall forests of mature pine and spruce. Clean, free-flowing rivers teeming with fish and birds. The sound of water rushing over stone.

A salmon river is a nice place to call home.

Notes

CHAPTER 1: THE PRESIDENT'S SALMON

1 **Claude and Rosemae Westfall:** C.Z. Westfall, interview with the author, October 2010. See also George Bush's kissin' cousin, political briefs. *The New York Times*, 26 May 1992.
2 **Hundreds of thousands of salmon:** Salmon migrating upstream are counted at fish passage facilities, usually on the first/most seaward dam and sometimes at later dams. In the case of the Penobscot in 1992, a fish ladder at the Veazie Dam directed salmon into a trap, where they were counted, measured, tagged, etc., on a daily basis by the Maine Atlantic Sea-Run Salmon Commission. Total population numbers, estimated based on these count data, are reported annually by the U.S. Atlantic Salmon Assessment Committee (USASAC). Estimates of the total historic Atlantic salmon population returning to U.S. rivers range from 300,000-500,000; the best estimate for the Penobscot is an annual run of 100,000 salmon; Saunders et al. (2006).
3 **nine and a half pounds:** This figure from Claude Westfall and confirmed by *Bangor Daily News* photo and caption in Claude Westfall's files; Baum (1997) says the salmon was 7.5 pounds.
4 **In the entranceway:** Whitcomb and Whitcomb (2000).
4 **Bush joined a long line of fishing presidents:** A. Phillips. Another Bush who's hooked on fishing. *The Washington Post*, 14 January 2001, p. D17; see also Mares (1999).
4 **"I've fished a couple of Oregon rivers . . .":** Mares (1999), citing 1989 interview in *Fly Rod & Reel* magazine by Jim Merritt.
4 **chow down on beef jerky:** Whitcomb and Whitcomb (2000).
4 **The human relationship with salmon:** Aas et al. (2011); Verspoor et al. (2007); World Heritage Center. Rock art of Alta, http://whc.unesco.org/en/list/352.
4 **The salmon's effortless:** Atlantic Salmon Federation (1997). About Atlantic salmon: cultural history, webpage (no longer accessible).

5 **In the first century:** Pliny the Elder. Historis naturalis, Book 9, Chs. 18 and 32.
5 **In the eleventh century:** "There was no lack of salmon either in the river or in the lake, and it was bigger salmon than they had ever seen." E. Haugen (1942). *Voyages to Vinland*. New York: Alfred A. Knopf, archive.org/details/voyagestovinland013593mbp.
5 **The indigenous Beothuks:** J. P. Howley (1915). *The Beothuks or Red Indians*. Cambridge University Press, https://archive.org/details/beothucksorredin00howl; Dunfield (1985).
5 **The salmon is associated with prophecy:** Atlantic Salmon Federation. 1997. About Atlantic salmon: cultural history, webpage (no longer accessible).
5 **The reddish flesh:** Smith (2004). Pigmented flesh, a trait unique to four of the salmonid fishes, is caused by their capacity to accumulate dietary carotenoids in muscle tissue. In the case of Atlantic salmon, a primary source is the zooplankton *Calanus finmarchicus*. Rajasingh, H., D.I. Våge, S.A. Pavey, and S.W. Omholt (2007). Why are salmonids pink? *Canadian Journal of Fisheries and Aquatic Sciences* 64:1614-1627.
5 **"The Salmon is accounted the King":** Walton (1939).
5 **But by the early 1990s:** Claude's fish was not the first fish of 1992—more than twenty had been caught before his—but as he said, "Not everyone wanted to give their fish way."
6 **"The Endangered Species Act was intended as a shield":** M. Wines. Bush, in Far West, sides with loggers. *The New York Times*, 15 September 1992, p. A25.
6 **a broken law that would not stand:** Schneider, K. U.S. to act faster on saving species. *The New York Times*, 16 December 1992, p. A1.
6 **When the Act threatened to block a dam:** Shabecoff (1993). It is unclear what specific case Shabecoff refers to; possibly the fountain darter and water withdrawal from the Edwards Aquifer or the Stacy Reservoir/Concho water snake.
7 **Owls would be captured:** T Egan. Strongest U.S. environment law may become endangered species. *The New York Times*, 26 May 1992, p. A1.
7 **On the Penobscot, industry put forth:** T.F. Darin (2000). Designating critical habitat under the Endangered Species Act: habitat protection versus agency discretion. 24 *The Harvard Environmental Law Review* 209.

CHAPTER 2: SEBOOMOOK

10 **The ice extended far out into the North Atlantic:** H.W. Borns et al. (2004). The deglaciation of Maine, U.S.A., pp. 89-109 in *Quaternary Glaciations – Extent and Chronology*, Part II (J. Ehlers and P.L. Gibbard, eds.) Elsevier.
10 **Salmonids are among the most northern of freshwater fish:** M. Kinnison, interview with the author, 10 June 2014. See also Bernatchez, L., and C.C. Wilson (1998). Comparative phylogeography of Nearctic and Palearctic fishes. *Molecular Ecology* 7:431-452.
10 **Salmon that had traveled into deeper basins:** O.K. Berg (1985). The formation of non-anadromous populations of Atlantic salmon, *Salmo salar* L. in Europe. *Journal of Fish Biology* 27:805-815; King, T.L. et al. 2001. Population structure of Atlantic salmon (*Salmo salar* L.): a range-wide perspective from microsatellite DNA variation. *Molecular Ecology* 10:807-821; Caldwell, D.W., and L.S. Hanson (2003). The isolation of landlocked salmon in Maine and Quebec near the end of the last ice age (abstract). Annual Meeting of the Geological Society of America, 5 November 2003, Seattle, WA.
10 **drainage system of the emerging Penobscot:** During the period of adjustment after the Ice Age, water flows in the Penobscot were 25 percent greater than at present, in part because Moosehead Lake, now the headwaters of the Kennebec River, drained into the Penobscot for a relatively brief period. Ice is heavy, and its weight pushes down on the landscape, forming

a bulge at the glacier's edge. When the ice melted and the weight was removed between 10,000 and 9,000 years ago, the bulge rippled through the landscape, like a wrinkle in a carpet, from southeast to northwest. This event tilted the land, interrupting the developing drainage patterns and tipping Moosehead Lake into the West Branch Penobscot River at Northwest Carry, just downstream from Seboomook. As the bulge passed, the land relaxed and tilted back south toward the sea, shunting the outlet of Moosehead Lake into its current drainage of the Kennebec and sending water tumbling over new channels of fresh rock. Balco, G., D.F. Belknap, and J.T. Kelley. 1998. Glacioisostasy and lake-level change at Moosehead Lake, Maine. *Quaternary Research* 49:157-170; Kelley and Sanger (2003).

11 **People moved in from the south and west:** Bourque (2001). For chronology, see also Kelley and Sanger (2003); Sanger, D. 2006. An introduction to the Archaic of the Maritime Peninsula: the view from central Maine, pp. 221-252 in Sanger and Renouf (2006).

11 **basins filled with rain:** H. Almquist-Jacobson and D. Sanger (1999). Paleogeographic changes in wetland and upland environments in the Milford drainage basin of central Maine in relation to Holocene human settlement history, pp. 177-190 in *Current Northeast Paleoethnobotany* (John Hart, ed.). Albany, NY: New York State Museum Bulletin No. 494; Davis, R.B., and D.S. Anderson (2001). Classification and distribution of freshwater peatlands in Maine. *Northeastern Naturalist* 8.1:1-50.

11 **The people walked:** Sanger and Renouf (2006), p. 235; also Bourque (2001), p. 44.

11 **In the graves they placed finely crafted plummets:** "Does the plummet have precisely designated associations in ritual contexts, or is it a fishing weight that was deposited with line or net for use in the afterlife?" asked Brian Robinson, in Sanger and Renouf (2006), p. 358. Maine anthropologist David Sanger cautions against using the term "Red Paint People" to describe the red ochre burial tradition, because there is no evidence that they were a distinct race, a common misrepresentation. D. Sanger (2000). "Red Paint People" and other myths of Maine archaeology. *Maine History* 39:145-168.

11 **Along the high cliffs of Seboomook:** Abbe Museum. 2012. Layers of time online exhibit: Seboomook Lake. Bar Harbor, ME, http://abbemuseum.org/exhibits/online/layers-of-time/seboomook-lake.html; see also R. Cole-Will and R.T. Will (2006). Final Report: 2004 Abbe Museum Fieldschool at Site 130.35, Seboomook Lake. Bar Harbor, ME: Abbe Museum.

12 **The Penobscot River flows through hundreds of occupation sites:** A. Kelley, interview with the author, 2 May 2011.

12 **The oldest salmon bones:** B.S. Robinson G.L. Jacobson, M.G. Yates, A.E. Spiess, and E.R. Cowie (2009). Atlantic salmon, archaeology and climate change in New England. *Journal of Archaeological Science* 36(10):2184-2191.

12 **They have always been here:** John Banks, personal communication with the author, February 2007. See also B. Phillips (2006). A River Runs Through Us. Augusta, ME: Penobscot River Restoration Trust; W. Pelletier (1974). For every North American Indian that begins to disappear, I also begin to disappear, from *Canada's Indians: Contemporary Conflicts* (J.S. Frideres, editor) Scarborough, Ontario: Prentice-Hall.

12 **Great white swans came from the east:** With their great billowing white sails and gliding motion, sailing ships looked like swans. In 1400-1500, fishing vessels from France, Spain, and Portugal had been working the waters of the Northwest Atlantic, trading with natives, who thus were well aware of (and not at first alarmed by) strangers by the time Europeans began settling the Penobscot Valley; Kolodny (2012).

12 **The natives pointed:** Cultural contact between Europeans and Native American groups often included native peoples communicating their knowledge of land to the newcomers. This transmitted knowledge sometimes took the form of gestures, words, a stick, finger, birchbark, or ink on paper; Pawling (2007), p. 6. "Maps helped to render Indian peoples

invisible in their own land, as cartographers contrived to promote a durable myth of an empty frontier," J.B. Harley (1994). New England Cartography and the Native Americans, pp. 287-313 in Baker et al. (1994).

12 **The strangers borrowed local words:** Place-names are from F.H. Eckstorm (1978). Indian Place Names of the Penobscot Valley and the Maine Coast. Orono, ME: University of Maine Press. Naming Places the Penobscot Way (special section), *Bangor Daily News*, 23-24 September 2006. See also http://www.wabanakitrails.org/nomenclature/.

13 **They hunted beaver:** According to tribal historian James Francis, the decimation of the beaver population removed a prized resource and drained the landscape, reducing the Penobscots's ability to travel through the watershed in the way they were accustomed, via canoe (Research Café, 20 April 2006, University of Maine). The fog was referenced by Nicolar (1893), and according to Kolodny (2012) likely refers to ash from a volcanic eruption that coincided with the increasing appearance of Europeans.

13 **the native was an enemy:** The Wabanaki generally sided with the French and fought on the side of French colonists throughout numerous wars between the French and English. When the English gained control of New England, the Wabanaki were considered enemies (although they fought alongside English colonists against Britain during the Revolutionary War). The Penobscot region was a refuge for those who had given up trying to live near the growing European settlements to the south and west.

13 **they put a bounty on his head:** Abbe Museum (2012). Phips Proclamation 1755, Wabanaki Timeline. Bar Harbor, ME: Abbe Museum, http://abbemuseum.org/research/wabanaki/timeline/proclamation.html.

13 **they signed a treaty:** Native Americans first signed a peace treaty with the English in 1676; between then and 1752, they signed 13 treaties. After the American Revolution, the Penobscots signed treaties with Massachusetts defining the limits of their lands (1786); reducing their territory (1796); and permitting road construction and use of waterways within their territory (1818). In 1820, Maine adopted the 1818 treaty; N.N. Smith (2005). The rebirth of a Nation? A chapter in Penobscot history, pp. 407-423 in Papers of the Algonquian Conference, University of Manitoba.

13 **The falls at Old Town:** Pawling (2007); Prins (1994).

13 **Before there was a river:** B.L. Dana (2002). Speech to the Maine Legislature, 6 April 2002, reprinted by *Indian Country Today*, http://indiancountrytodaymedianetwork.com/2002/04/06/penobscot-leader-gives-historic-speech-maine-state-legislature-87641. Barry Dana also said in his speech, "We pray for the return of the salmon so our subsistence rights can be realized."

14 **Driven by thirst:** Prins (1994), p. 11; Speck (1997), p. 82.

14 **a way to eliminate Great Britain's national debt:** Dunfield (1985).

14 **Hawley Emerson of Phippsburg:** description of commercial weir fishing from Atkins (1887).

15 **"to prevent an evil so great":** Petition of Joseph Pease, Franwook Sabion, Captain Nichola, and Joseph Loling, to Massachusetts's governor and General Court, Boston, 24 September 1801, 1801 Senate Document #2838, 1802 unpassed legislation, MA State Archives, cited by Pawling (2007).

15 **But the Penobscots were few:** Even before the Pilgrims first landed at Plymouth in 1620, between 1616 and 1619, during a period labeled by the English as the "Great Dying," European diseases killed 75 percent of natives in coastal areas from Cape Cod to Penobscot Bay, Kolodny (2007).

15 **All salmon and other members of the family Salmonidae:** Salmon evolution description from Gross, M.R., R.M. Coleman, and R.M. McDowall (1988). Aquatic productivity and

the evolution of diadromous fish migration. *Science* 239:1291-1293; Stearley. 1992. Stearley, R. F., and G.R. Smith (1993). Phylogeny of the Pacific trouts and salmons *(Oncorhynchus)* and genera of the family Salmonidae. *American Fisheries Society Transactions* 122:1-33; Kinnison, M.T., and A.P. Hendry (2004). From macro- to micro-evolution: tempo and mode in salmonid evolution, pp. 208-231 in Hendry and Stearns (2004); S.C. Stearns and A.P. Hendry (2004). The salmonid contribution to key issues in evolution, pp. 3-19 in Hendry and Stearns (2004); Verspoor et al. (2007); R.S. Waples, G.R. Pess, and T. Beechie (2008). Evolutionary history of Pacific salmon in dynamic environments. *Evolutionary Applications* 1(2):189-206; A. Crete-Lafreniere, L.K. Weir, and L. Bernatchez (2012). Framing the Salmonidae family phylogenetic portrait: a more complete picture from increased taxon sampling. *PLOS One* 7(10):e46662; J.L. Nielsen, G.T. Ruggerone, and C.E. Zimmerman (2013). Adaptive strategies and life history characteristics in a warming climate: salmon in the Arctic? *Environmental Biology of Fishes* 96:10-11, 1187.
16 **Surviving salmon took refuge:** In Europe, which did not have as great an extent of ice cover as America, the refuge areas include the Iberian Peninsula (those Spanish caves), the North Sea, and the ice lakes east of the Baltic Sea.
16 **Post-glacial pioneers:** M. Kinnison, interview with the author, 10 June 2014.

CHAPTER 3: A TRADITION BEGINS

19 **Winter still held:** Ice moved a little, *Bangor Daily News*, 2 April 1912. For description of salmon fishing and account of presidential recipients, see Baum (1997); Smythe (1953).
19 **Anderson had immigrated to Maine:** Jagels (2010).
20 **Electricity had come to Bangor:** Reilly, W.E. Electricity made Bangor 'metropolitan.' *Bangor Daily News*, 7 February 2011, p. D1.
20 **In the middle of the channel:** See map in 1912 *Bangor City Directory*, Bangor Room, Bangor Public Library.
20 **Salmon sustained Benedict Arnold's Revolutionary War troops:** Watson, W.C. (1876). The salmon of Lake Champlain and its tributaries, pp. 531-540 in *Report of the Commissioner for 1873-1874 and 1874-1875*, Part III, U.S. Commission of Fish and Fisheries. Washington, DC: Government Printing Office.
20 **dinner on the first Independence Day:** *American Heritage Cookbook and Illustrated History of American Eating & Drinking* (1964). New York: American Heritage Publishing Company, p. 406; Cannon and Brooks (1968); Standish, M. (1984). *The Best of Marjorie Standish Seafood Recipes*. Portland, ME: Gannett Books, p. 48. See also *Bangor Daily Commercial*, 10 April 2013, p. 9.
20 **"No fish of its magnitude":** Baird, S.F. (1874). *Report of the Commissioner for 1872 and 1873*, Part II, U.S. Commission of Fish and Fisheries. Washington, DC: Government Printing Office, p. 62.
20 **"everywhere-popular":** Nickerson, A.R. (1905). *28th Report of the Commissioner of Sea and Shore Fisheries of the State of Maine*. Augusta, ME: Kennebec Journal Print; Nickerson, A.R. (1907). *29th Report of the Commissioner of Sea and Shore Fisheries of the State of Maine*. Augusta, ME: Kennebec Journal Print.
20 **"Penobscot Salmon, Sauce Hollandaise":** See, for example, the April 1884 banquet menu for the Champlain Society at Harvard University, Champlain Society archives, Mount Desert Island Historical Society.
20 **"Penobscot River salmon are the best":** Farmer, F.M. (1896). *Boston Cooking School Cook Book*. Boston: Little, Brown and Company, p. 138.

21 **"each with a setter dog"**: Smythe (1953). Dues were $2 in 1887 and went up to $10 in the early twentieth century. Maine fish warden Tom Allen helped to organize the club (*Forest and Stream*, 5 May 1887).

21 **"This explanation is perhaps"**: Dymond (1963).

21 **Fly fishing was practiced on the Penobscot**: "The migratory salmon of Maine were deemed sullen, wary, capricious, reluctant to take the fly, although tourist-anglers fished for land-locked salmon in Grand Lake (with the help of Passamaquoddy Indian guides)" Atkins (1874). See also Wells (1886); Atkins (1887); Kendall (1935); Netboy (1968, 1974); Schullery (1999).

22 **He would not cast across a pool**: *Forest and Stream*, 5 May 1887.

22 **Karl decided the other salmon**: Fine salmon for President Taft. *Bangor Daily News*, 3 April 1912, p. 1; Salmon for Taft. *Bangor Daily Commercial*, 2 April 1912, p. 1.

22 **Bangor delegates to the Republican State Convention**: Bangor delegates all declare for Mr. Taft. *Bangor Daily Commercial*, 3 April 1912, p. 1.

22 **On the West Coast**: Harrison, J. (2008). First-salmon ceremony. Columbia River History Project. Portland, OR: Northwest Power and Conservation Council, http://www.nwcouncil.org/history/firstsalmonceremony.asp.

23 **The Ainu people**: Roche, J., and M. McHutchinson, eds. (1998). *First Fish First People: Salmon Tales of the North Pacific Rim*. Vancouver: UBC Press.

23 **Servants of the Imperial household**: Netboy (1974).

23 **The site of London's Westminster Abbey**: Hall, J. (2011). The Very Reverend Dr. John Hall, Sermon given at Sung Eucharist at Petertide 2011, Westminster Abbey, London, 29th June 2011 at 5:00 pm, http://www.westminster-abbey.org/worship/sermons/2011/june/sermon-given-at-sung-eucharist-at-petertide-2011; R. McGeorge, London Fishmongers' Company, email to the author, 12 May 2014.

23 **Knights celebrating a jousting victory**: Netboy (1974).

24 **The presidential salmon custom maintained**: Tennahill, R. (1988). *Food in History*. New York: Three Rivers Press, p. 80; Smith (2004).

24 **in a crate with straw and ice**: New Englanders had been shipping fish and other foods on ice via train since the 1830s, and Maine was the leading ice-producing state, exporting three million tons in 1890; Kurlansky (2012).

24 **received by housekeeper Elizabeth Jaffray and given to Alice Howard**: based on Seale (1986).

24 **Taft was the gourmand**: Taft biographical information from Coletta, P.E. (1973). *The Presidency of William Howard Taft*. Lawrence, KS: University Press of Kansas.

24 **President Taft was in the study**: Taft, W.H. (1912). *Economy and Efficiency in the Government Service: Message of the President of the United States Transmitting Reports of the Commission on Economy and Efficiency*, Document No. 670, 4 April 1912. Washington, DC: Government Printing Office; *The Need for a National Budget: Message from the President of the United States Transmitting Report of the Commission on Economy and Efficiency on the Subject of a Need for a National Budget*, Document 854, 27 June 1912. Washington, DC: Government Printing Office.

24 **"efficiency craze"**: Hays (1959).

25 **the rift between Taft and Roosevelt**: For an intimate account of the relationship between Taft and Roosevelt, see Manners, W. (1969). *TR and Will: A Friendship that Split the Republican Party*. New York, NY: Harcourt, Brace & World, Inc.

25 **signing numerous bills**: Hays (1959).

26 **the first Mill Act**: *Laws of the State of Maine*, volume I, 1821, chapter 45. Brunswick, ME: J. Griffin.

26 **the only thing mill owners couldn't do:** Babb (1912).
26 **"as dense as that found in the manufacturing districts":** Poor, J.A., A.D. Lockwood, and H. Hamlin. (1868). *Report of the Commissioners of the Hydrographic Survey of the State of Maine 1867.* Augusta: Owen & Nash.
26 **prompting investors to promote their fledgling industry:** Davies (1972); Schurr and Netschert (1973); Croul, R. Street railway was answer to excess of electric power. *Bangor Daily News,* 28 June 1984, p. C4.
27 **"offers grand opportunities to the capitalist":** Boardman, W.H. (1901). Development of water power of the Penobscot River at Vinal's Landing, Webster, B.S. thesis, University of Maine.
27 **"In traveling along the lower part of the Penobscot River":** Cole, W.L. (1903). A preliminary investigation of the conditions for establishing a waterpower at Birch Island Rips near Greenbush, Maine, B.S. thesis, University of Maine.
27 **"wild land" known only to the lumberman and the sportsman:** Wells (1868).
27 **Holding total faith that demand would follow supply:** "The necessity for the immediate development of this power is not apparent but in consideration of the growth of the electrical industry and demand for power [by household appliances] it is reasonable to expect that this power will be necessary before many years." Mullen, J.N. and C.R. Archer (1923). The development of hydro-electric power at Mattaceunk Rips Mattawamkeag Maine, B.S. thesis, University of Maine.
27 **"The old dam is not only out of date":** Bangor Power Co to build new dam at Veazie. *Bangor Daily Commercial,* 5 April 1912, p. 14.
27 **cook their salmon on an electric stove:** Reilly, W.E. Electricity made Bangor "metropolitan." *Bangor Daily News,* 7 February 2011, p. D1.
28 **the fish had gotten smaller:** As with other fish, harvesting pressure on salmon tends to make the population shift toward smaller and younger individuals. Salmon biologists don't believe that grilse (salmon that return after only one winter at sea and are generally less than twenty-seven inches) were a large part of New England salmon runs; the large, multi-winter fish that used to run early in the spring had disappeared by 1940, after which grilse made up 20-50 percent of returning adults (J. Kocik, email to the author, 28 October 2014). Today, one sea winter fish are fewer than 25 percent of returning adults (R. Spencer, email to the author, 28 October 2014).
28 **"exceedingly limited":** Smith (1897). See also *Forest and Stream,* 5 December 1889, 13 August 1910, and 24 April 1890; Reilly, W.E. In 1908, Bangor fishermen fretted over salmon decline. *Bangor Daily News,* 7 April 2008, p. C6.
28 **Rapidly evolving art forms:** Lears, T.J. 1981. *No Place of Grace: Antimodernism and the Transformation of American Culture, 1880-1920.* New York, NY: Pantheon.
28 **The Arts and Crafts movement:** Ibid.; Holm (2005).
28 **"correct" fishing:** Reiger (2001), p. 52.
29 **perfected by Hiram Lewis Leonard:** Butler and Taylor (1992).
29 **To lure and land an Atlantic salmon:** Wells (1886).
29 **"The lover of the 'gentle craft'":** Hamlin, A.C. (1869). On the salmon of Maine. *Lipincott's Magazine,* May 1869, reprinted in Baird, S.F. 1874. Report of the Commissioner for 1872-1873, Part II, U.S. Commission of Fish and Fisheries. Washington, DC: Government Printing Office.
29 **people who needed the food:** Judd (1997), pp. 197-209; Schullery (1999).
29 **Traditional ways of acquiring fish for food:** Judd (1997), pp. 206-208.
29 **Sport fishermen claimed:** Kendall (1935).
29 **The native Penobscots had developed many ways to catch fish:** Speck (1997).

30 **From a ledge atop the falls:** Speck (1997), pp. 82-91; Denys, N. (1908). *The Description and Natural History of the Coasts of North America (Acadia)*, Translated and Edited, with a Memoir of the Author, Collateral Documents, and a Reprint of the Original, by William F. Ganong. Toronto: The Champlain Society, pp. 436-437. The Mi'kmaqs of Maine and Atlantic Canada called salmon *plamu*; Maliseets and Passamaquoddies, *polam*. Narragansett Indians, *mishquammauquock*. Penobscots fished from the ledges at Bangor, Veazie, and Old Town.

30 **in 1912 they reinstated an 1883 law prohibiting salmon fishing:** *Laws of Maine*, (1913), 76th Legislature Including Acts and Resolves of the Special Session held in 1912, Chapter 206, pp. 257-258. The University of Maine Hudson Museum, and the Maine Memory Network, have a record of a fish spear, ca. 1900. The description for this material states "Salmon spearing was prohibited by state law in 1912." A review of legislative activity in 1911-1913 revealed only a consolidation of fish and game laws. A review of earlier laws in the Maine State Law and Legislative Library revealed that the spearing prohibition first appeared in 1878, but the law pertaining to spearing specifically for "sea salmon" first appeared in 1883.

30 **"grapnel, spear, gaff with more than one prong.":** Laws of Maine, 1913, 76th Legislature Including Acts and Resolves of the Special Session held in 1912, Chapter 206, pp. 257-258. For an account of similar treatment of native fishermen in the Canadian Maritimes, see Parenteau, B. (1998). 'Care, control and supervision': Native people in the Canadian Atlantic salmon fishery, 1867-1900. *The Canadian Historical Review* 79(1):1-35.

30 **"It is claimed with some show of justice":** Norris, T. (1865). *American Angler's Book*. Philadelphia, PA: E.H. Butler & Co.

31 **"There is not a cubic foot of water in the whole country":** *Forest and Stream*, 16 June 1892; also Reiger (2001), p. 46.

31 **Many believed that indigenous cultures and beliefs were dying out:** Holm (2005), p. 27.

31 **Poverty, cholera, influenza, and turberculosis had decimated their population:** Kolodny (2007), p. 22.

31 **the farming encouraged by the government:** Prins (1994), p. 102; MacDougall (1995). Crop subsidies to the Penobscots began in 1839. The Penobscots were criticized as impatient and lazy—but they were never farmers. The Meso-American agriculture of corn, squash, and beans, the three sisters, stopped at the Kennebec River to the west; to the east, the climate and landscape did not favor horticulture, Prins (1994), p. 102. The Penobscots were not subject to federal Indian policies, although they were exposed to the same ideas that forged federal Indian policies, like the idea that Indians should take up agriculture in place of fishing, hunting, and trapping as a way of "assimilating." As incentive, they received payment for crops; in 1912, the Penobscots received payments for 1,830 bushels of potatoes, 176 bushels of beans, and 782 bushels of oats.

31 **Isolated in their main village on Indian Island:** Kolodny (2007), p. 16-17.

32 **poached whole, bedecked with curly parsley:** S. Oliver, email to the author, 3 December 2010. Cannon and Brooks (1968) reported that Taft did not like eggs, so an egg sauce was unlikely.

CHAPTER 4: CHESUNCOOK

34 **"beauty strips":** Lansky (1992), p. 6.

34 **Charles Atkins attached aluminum tags:** Baird, S. (1876). *Report of the Commissioner for 1873-74 and 1874-5*, Part III, U.S. Commission of Fish and Fisheries. Washington, DC: Government Printing Office.

35 **They are guided, too, by smell:** That salmon use smell was documented in the 1960s by Arthur Hasler and Warren Wisby of University of Wisconsin-Madison. There is also evidence of a genetic basis—salmonids have a genetic tendency to return to a specific natal site/habitat type even if they did not imprint on that site, though imprinting is a stronger influence. Hendry, A.P., V. Castric, M.T. Kinnison, and T.P. Quinn. The evolution of philopatry and dispersal: homing versus straying in salmonids, pp. 52-91 in Hendry and Stearns (2004); also Thorstad et al. (2011).

35 **Only the strongest succeed:** Crie, H.D. (1928). *Fifth Biennial Report of the Commission of Sea and Shore Fisheries of the State of Maine*. Rockland, ME.

35 **searching for the one place they've known:** Thorstad et al. (2011). While scientists know that upstream migration is affected by water temperature and discharge, they still can't reliably predict the springtime arrival of a salmon.

35 **only two out of every hundred Penobscot River fish stray from their home stream:** Baum, E., and R.C. Spencer (1990). Homing of adult Atlantic salmon released as hatchery-reared smolts in Maine rivers. Working paper 1990-17. Working Group on North Atlantic Salmon, International Council for the Exploration of the Sea.

36 **For a long time, Europeans believed:** Netboy (1974), p. 92.

36 **"one of the wildest, most picturesque spots":** F.C.P. Down the Penobscot by Canoe. *Forest and Stream* 11(20):398, 19 December 1978.

36 **"The water foams and hisses":** Hubbard, L.L. (1879). *Summer Vacations at Moosehead Lake and Vicinity*. Boston: A. Williams and Company.

36 **Historical reports suggest:** In 1866, Edward G. Masterman of Greenville told Maine Fish Commissioner Charles Atkins that "in August and September of any year he can catch young salmon from six to twelve inches long at Ripogenus Falls." And Joel M. Lane reported that one blustering night in November 1838, he speared nine large salmon on Ripogenus falls. Joe Francis, Atkins' Penobscot guide, saw them in Penobscot Brook, the headwaters of the West Branch. Foster and Atkins (1868, 1871).

36 **all good breeding grounds for salmon:** Atkins (1874).

36 **Perhaps more than any other part of the Penobscot watershed:** Smith (1972), p. 79. The Penobscot Log-Driving Company began taking logs from the woods around Chesuncook in 1856.

37 **"King Pines":** Smith (1972), citing William Williamson. The first cargo shipped back to Britain from the American colonies was a load of masts in 1609; the country had been chronically short of timber since the 1500s, and had been using Riga firs from the Baltic region. A main mast on a sailing ship required a straight tree 120 feet long and forty inches in diameter. According to Wilkins (1932), the search for "Mast Pine" and the King's broad arrow policy did not reach as far east as the Penobscot.

37 **their scent alerting weary sailors:** Burnaby, Rev. A. (1968). *Travels through the Middle Settlements in North America . . . 1759-1760*. Ithaca, NY: Cornell University Press, p. 70, cited in Dunfield (1985).

37 **"greatest scene":** Wilkins (1932). The first sawmill in New England was built in York, Maine, in 1623, Pike (1967). Sawmills were built on tributaries of the lower Penobscot including Sedgeunkedunk Stream, Eaton Brook, and Penjajawoc Stream (1772), Orland River and Mill Creek (1773), Souadabscook Stream (1782), Marsh Stream (1785), and Kenduskeag Stream (1795). The first mills on the mainstem of the Penobscot included the series begun on the Stillwater Branch at Orono in 1778 and on the Penobscot at Old Town in

1798. Boardman (1960) states that the first mill in 1771 was at the mouth of Penjajawoc Stream, and the first timber yard at the mouth of Kenduskeag.

37 **"timber cruisers":** Pike (1967). In 1870 Maine gave away its last 800,000 acres of forest land to the European and North American Railway.

38 **"all at once the apparently solid surface would begin to creak and settle":** Pike (1967).

38 **Logs left behind:** McLeod (1978), chapter 4, pp. 10-30.

38 **dug channels between lakes:** There was talk of shunting the Penobscot into Moosehead Lake at Northeast Carry near Seboomook. In 1841, Shepherd Boody built a dam that raised the water level in Chamberlain Lake, just north of Chesuncook, and dug a channel to direct the water into Telos and Webster lakes. The Telos Cut redirected water that naturally would have flowed into the Allagash River and then the St. John to the Penobscot, so that northern Maine lumber would not be shipped to Canada. This alteration added 249 square miles to the Penobscot watershed, enlarging it by almost 3 percent.

39 **The -*cook* ending:** Cook (2007); J.E. Francis, interview with the author, 11 July 2014.

39 **promise not to draw any maps:** Eckstorm, F.H. (1926). History of the Chadwick Survey. *Sprague's Journal of Maine History* 14(2):62-89.

40 **"a fine sheet of water":** Greenleaf, M. (1829). *A Survey of the State of Maine in Reference to its Geographical Features, Statistics, and Political Economy*. Augusta, ME: Maine State Museum, reprint 1970.

40 **Treat drew the river:** Pawling (2007). Although Wabanaki peoples in Maine were among the first Native groups on the eastern seaboard of North America to experience European contact, the economic and political uncertainties of a borderland region slowed European settlement of the area. According to Treat, settlers had only reached as far up the Penobscot as Howland.

40 **Treat competed with the Penobscots:** Resolve on the Petition of Joseph Treat, chapter CLV, 27 February 1813, Resolves of the General Court of the Commonwealth of Massachusetts, Passed at the Session, in October 1812, and January 1813. Boston: Russell and Cutler, cited in Pawling (2007).

40 **One of the most difficult portages:** Jackson, C. (1838). *Second Annual Report on the Geology of the Public Lands Belonging to the Two States of Maine and Massachusetts*. Augusta, ME: Luther Severance.

40 **In times of high water:** Cook, D.S. (1984). Preliminary Assessment of Archaeological Potential: West Branch Penobscot below Ripogenus Dam, Appendix in Great Northern Paper's Application for License for Big "A" Hydroelectric Project, FERC No. 3779, Volume VIII, Exhibit E, March 1984.

41 **Treat had intended to go up the West Branch:** Pawling (2007).

41 **site of the first major dam:** Ibid., pp. 121-133. Of course, the Natives traveled upstream and downstream, as long as there was water they could move throughout the entire region. Hempstead (1931) dates the dam to 1840.

41 **The first dam to stretch all the way:** R. Judd, personal communication, October 2014; Atkins (1887); Holbrook, C. How has fish passage on the Penobscot River changed with the development of dams and fishways since 1820? Unpublished manuscript prepared for NOAA Fisheries, Orono, ME. The Milford Dam was the first to stretch across the full width of the river, but Veazie Dam was the first to completely block fish passage.

41 **"superior and controlling use of the waters of the West Branch":** *Charles W. Mullen v. Penobscot Log-Driving Company*, 20 September 1897, Hamlin, C. Cases argued and determined in the Supreme Judicial Court of Maine 1897. *Maine Reports* 90:555-570; see also Pike (1967).

41 **including many Penobscots:** Bourque (2001), p. 222. Also 1910 census, viewed on display at Abbe Museum.
42 **The lumber trade hit its peak in 1872:** Boardman (1960); Pike (1967); Hale (2005).
42 **When Henry David Thoreau visited:** Hale (2005).
42 **"In many places, after picking out the large rocks":** Krohn, W.B., ed. (2005). *Manly Hardy (1832-1910): The Life and Writing of a Maine Fur-Buyer, Hunter, and Naturalist, Northeast Folklore* 38. Orono, ME: Maine Folklife Center.
42 **salmon could surmount crude fishways:** Foster and Atkins (1869).
43 **America had lost more than half its Atlantic salmon habitat:** Atkins (1874); Goode (1884). "The Penobscot is now the only stream on the Atlantic Coast of the United States having an important run of salmon," Smith, H.M. (1898). Salmon Fishery of Penobscot Bay and River, Report of the Division of Statistics and Methods of the Fisheries, pp. 130-131 in *Report of the Commissioner for the Year Ending June 30, 1897*, Part XXIII, U.S. Commission of Fish and Fisheries. Washington, DC: Government Printing Office. The Penobscot River "is the only New England stream that now has a run of salmon, and even the Penobscot maintains its salmon in the face of serious physical and other handicaps," Smith, H.M. (1921). *Report of the U.S. Commissioner of Fisheries for the Fiscal Year 1920*. Washington, DC: Government Printing Office, p. 44; Kocik and Brown (2002).
43 **hurled themselves upon the piers at Chesuncook:** Pike (1967). The principle food of logging camps, however, was pickled beef and dried cod, sourdough biscuits, and flapjacks. Beans, pork, molasses and gingerbread, black tea, and potatoes became staples after 1850.
43 **Manly Hardy reported taking salmon:** Atkins (1874); Stilwell, E.M., and H.O. Stanley (1873). *Seventh Report of the Commissioners of Fisheries of the State of Maine*. Augusta: Spraque, Owen & Nash.
43 **River drivers noticed:** Baird, S.F. (1880). *Report of the Commissioner for 1878*, Part VI, U.S. Commission of Fish and Fisheries. Washington, DC: Government Printing Office, p. 925; Hubbard, L.L. (1879). *Summer Vacations at Moosehead Lake and Vicinity*. Boston: A. Williams and Company.
43 **"a very handsome fish":** Hatch, B.L., ed. 1940. Down the West Branch of the Penobscot, August 12-22, 1889, from the journal of Fannie Pearson Hardy (Mrs. Fannie Hardy Eckstorm). *Appalachia* December 1940.
43 **"Improvements" intensified:** Hempstead (1931); Hauger, N-N. That was our mill. *Bangor Daily News*, 7 October 2006, pp. 1, A5. Spruce was first cut on the Penobscot in 1845, Pike (1967); Wilkins (1932). The 1871 charter of the West Branch Dam and Improvement Company permitted the corporation to "erect dams, and make all other improvements on the West Branch of the Penobscot River and its tributaries, to facilitate the driving of logs down said river, from the head of Chesuncook Lake, and may construct steamers upon said lake, and upon the Pemadumcook and North Twin lakes, for the same purpose; and the said corporation may take land and materials necessary to construct their works . . . and the said corporation may flow contiguous lands . . ." *Private and Special Laws of the State of Maine*, 1871, chapter 564, pp. 555-557.
43 **Below Chesuncook Dam:** Hempstead (1931). Ripogenus dam was built in 1887.
44 **the corporation prevailed:** Ibid.
44 **In contrast, on the Mattawamkeag:** Army Corps of Engineers. (1930). Letter from the Secretary of War transmitting report from the chief of engineers on Penobscot River, ME, covering navigation, flood control, power development, and irrigation, House of Representatives Document 652, 71st Congress, Third Session.

45 **white ghosts against the verdure:** Hatch, B.L., ed. 1940. Down the West Branch of the Penobscot, August 12-22, 1889, from the journal of Fannie Pearson Hardy (Mrs. Fannie Hardy Eckstorm). *Appalachia*, December 1940.

CHAPTER 5: BIG BUSINESS

47 **A group of Bangor Democrats:** First salmon to the President. *Bangor Daily News*, 7 April 1916.
47 **Broiled shad, a founding fish:** McPhee, J. (2002). *The Founding Fish*. New York, NY: Farrar, Straus, and Giroux; Rhodes, S.R., ed. (1913). *The Economy Administration Cook Book*. Hammond, IN: W.B. Conkey Company, pp. 50-51, cited on Foodtimeline.org.
48 **preferred chicken salad:** Cannon and Brooks (1968).
48 **American attention had been focused on Europe:** Outwin, P.R. (2014). "Where hostilities are now in progress": Documents from the MHS WW1 Collection. Portland, ME: Maine Historical Society, mainehistory.wordpress.com/tag/world-war-i/.
48 **"Where has gone the once prosperous river of Penobscot?":** Reilly, W.E. Rumors plagued Maine at start of WWI. *Bangor Daily News*, 1 September 2014, pp. B1-B2.
48 **Per capita, Maine led the nation in developing hydroelectric power:** Brewster, R.O. (1927). Development and control of hydroelectric power, Special message by Governor Ralph O. Brewster, March 23, 1927, to the 83rd Legislature of the State of Maine in joint assembly. Augusta, ME.
48 **an "octopus" controlling much of the West Branch:** Beach (1993).
48 **all available water power should be used:** Public Utilities Commission (1918). *Special Water Power Investigation*. Lewiston, ME: Journal Printshop.
48 **the Fernald Law:** Beach (1993); Schurr, S.H., and B.C. Netschert (1960). *Energy in the American Economy, 1850-1975*. Baltimore: Johns Hopkins Press. See also Brewster, R.O. (1927). Development and control of hydroelectric power, Special message by Governor Ralph O. Brewster, March 23, 1927, to the 83rd Legislature of the State of Maine in joint assembly. Augusta, ME; Maine Development Commission (1929). The 1909 Fernald Law prohibiting power export remained on the books until 1955, when Governor Edmund Muskie deemed it impractical and unnecessary.
48 **"the measure of material civilization":** Pinchot, G. Great Power, message to the 83rd Legislature, 16 February 1926, Augusta, ME.
49 **"an immense concrete affair...":** Anonymous. (1907). Facts & figures in connection with the rebuilding and raising the Ripogenus Dam, Great Northern Paper Co. Papers, Special Files MS 210, Box 12, folders 12-16, Special Collections, University of Maine Fogler Library.
49 **Great Northern would give up rights to transmit power:** Beach (1993).
49 **justices in Maine's courts:** Spear, A.M. (1919). Answer to questions propounded to the court by order of the House of Representatives of February 20, 1919. 118 *Maine Reports* 523-543.
49 **permission to raise Chesuncook Lake another four feet:** McLeod (1978).
50 **thought the fears of John Muir:** Soden (1999).
50 **the act was disappointing and "feeble":** Link, A.S. (1926). *Wilson: The New Freedom*. Princeton, NJ: Princeton University of Press; Swain (1963); Wilson, W. (1966). *The Papers of Woodrow Wilson*, V. 45, p. 168. Princeton, NJ: Princeton University Press; Brown, P.W., and A.W. Buxton (1979). Preliminary analysis of legal obstacles and incentives to the development of low-head hydroelectric power in the Northeastern United States. Concord,

NH: Franklin Pierce Law Center; Fox, W.F. Jr. (1984). *Federal Regulation of Energy.* Colorado Springs, CO: Shepard's/McGraw-Hill; Federal Energy Regulatory Commission (1998). The Use and Regulation of a Renewable Resource, http://www.ferc.fed.us/hydro/docs/waterpwr.htm.

50 **The University of Maine became a training camp:** Soldier-artisans work and play at University of Maine. *Bangor Daily Commercial,* 29 June 1918.

50 **Meatless Tuesdays and Wheatless Wednesdays:** http://www.pbs.org/food/the-history-kitchen/history-meatless-mondays; http://www.whitehouse.gov/about/presidents/herberthoover.

51 **"salmon, either fresh or canned, equals round steak in protein":** Goudiss, C.H., and A.M. Goudiss (1918). *Foods That Will Win the War and How To Cook Them.* New York: Forecast Publishing Company.

51 **"masked the mess":** Bolster (2012).

51 **The lights came back on:** Allen (1931).

51 **"It is probable that Mr. Flanagan's salmon will grace the table":** Week-end fishermen fared very well. *Bangor Daily Commercial,* 4 April 1921; Mayor sends first salmon to President. *Bangor Daily News,* 4 April 1921; Season opens at salmon pool. *Bangor Daily Commercial,* 1 April 1921.

52 **Harding sent a thank-you letter:** White House guests eat Bangor salmon. *Bangor Daily News,* 11 April 1921.

52 **headed to Texas for some golf and tarpon fishing:** Mares (1999), citing *The New York Times,* 9 November 1920.

52 **The White House food was mediocre and Harding social events lacked luster:** Smith (1967); Cannon and Brooks (1968).

52 **Accessible to a fault:** Allen (1931).

52 **law enforcement agents seized:** *Bangor Daily Commercial,* 6 April 1921.

52 **Harding supported rapid resource development:** Swain (1963).

53 **"Disturbance of the status quo":** Allen (1931).

53 **Demand for many products, including fish, increased:** Cobb, J.M. (1917). The Pacific Salmon Fisheries, Appendix III, p. 146 in *Report of the U.S. Commissioner of Fisheries for the Fiscal Year 1916.* Washington, DC: Government Printing Office.

53 **Once scaled up for wartime needs:** Goldberg, R.A. 2003. *America in the Twenties.* New York: Syracuse University Press.

53 **Fannie Farmer had dropped:** Penobscot salmon were last highlighted in the 1918 edition of the *Boston Cooking School Cook Book.*

53 **The fishing industry soon spread north:** Woodcock, F.H., and W.R. Lewis (1938). *Canned Foods and the Canning Industry.* London: Sir Isaac Pitman and Son Ltd., London, cited in Dunfield (1985); Jarvis (1998).

53 **"a perfect web of nets":** O'Leary, W.M. (1981). The Maine Sea Fisheries, 1830-1890: The Rise and Fall of a Native Industry, Ph.D. dissertation, University of Maine; *Forest and Stream* 29(9), 22 Sept. 1887; Jordan, D.S. and C.H. Gilbert (1887). Part XIII, The Salmon Fishing and Canning Interests of the Pacific Coast, pp. 729-753 in *The Fisheries and Fishery Industries of the United States,* Section V, Volume I. Washington, DC: Government Printing Office.

54 **"Connoisseurs declare that the Penobscot salmon have a delicacy of taste":** Salmon for Taft. *Bangor Daily Commercial,* 2 April 1912.

54 **By 1916, more than eighty plants:** Smith, H.M. (1914). *Report of the U.S. Commissioner of Fisheries for the Fiscal Year 1913 with Appendices,* Bureau of Fisheries. Washington, DC: Government Printing Office, pp. 33-36; Cobb, J.M. (1917). The Pacific Salmon Fisher-

ies, Appendix III, p. 146 in *Report of the U.S. Commissioner of Fisheries for the Fiscal Year 1916*, Bureau of Fisheries. Washington, DC: Government Printing Office. Three salmon filled a case of cans, often packed by Chinese workers.
54 **readily available in Bangor markets:** Housewife reports on marketing trip. *Bangor Daily Commercial*, 19 July 1918.
54 **The U.S. Bureau of Fisheries:** The Office of Commissioner of Fish and Fisheries was designated as the Bureau of Fisheries in 1903 and Congress assigned it to the Department of Commerce.
54 **"met with universal favor":** Smith, H.M. (1919, 1920, 1921). *Report of U.S. Commissioner of Fisheries*, Bureau of Fisheries. Washington, DC: Government Printing Office; Northeast Fisheries Science Center. Fisheries Historical Page, http://www.nefsc.noaa.gov/history/ . Radcliffe, L. 1920. Fishery Industries of the United States, Report of the Division of Statistics and Methods of the Fisheries for 1918, Appendix X in *Report of the United States Commissioner of Fisheries*, Bureau of Fisheries. Washington, DC: Government Printing Office.
54 **"most common fish food":** Ford Educational Library (1922). Salmon Fishing (film), Industrial Geography of the United States, http://research.archives.gov/description/91179.
54 **New England fish processors responded to this competition:** Smith (2004).
54 **busy packing lobster, sardines, clams, and white fish:** Jarvis (1998), "At no time has the packing of salmon been of any importance on the U.S. Atlantic Coast," p. 183.
54 **Dozens of recipes for salmon loaf:** Oliver, S.L. (1995). *Saltwater Foodways*. Mystic, CT: Mystic Seaport Museum, Inc.
54 **"appeared to be hurtling toward collapse":** Taylor, J.E. III. (2004). Master of the seas? Herbert Hoover and the Western Fisheries. *Oregon Historical Society Quarterly* 105(1):40-61.
55 **Pacific canners inundated his office with fresh salmon:** Ibid.
55 **Hoover had re-embraced Hugh Smith's approach:** Ibid.
55 **Hatcheries and the bureaucracies that supported them:** Jenkins (2003); Taylor, J.E. III (1999). *Making Salmon: An Environmental History of the Northwest Fisheries Crisis*. Seattle, WA: University of Washington Press.
56 **"We all have a divine right to unlimited fish":** Hoover, H. (1930). *A Remedy for Disappearing Game Fishes*. New York: Huntington Press.
56 **J. Edward Canning caught the first salmon:** *Bangor Daily Commercial*, 2 April 1924. A second fish was sent to Coolidge by a private citizen later in the season. Mrs. Albert Whittier, from Bangor but living in Boston, bought Albert Fischer's twenty-pound salmon and shipped it as a gift to Coolidge; Penobscot salmon sent to President. *Bangor Daily Commercial*, 20 May 1924.
56 **"you ought to get lots of big salmon with this outfit":** Woman sent fish to Wilson. *Bangor Daily News*, 1 May 1982; Little Miss Sullivan and her big salmon. *Bangor Daily News*, 19 November 1925.
56 **Coolidge inspected the White House iceboxes:** Parrish (1992).
56 **challenged the White House cooks to come up with sauces:** Cannon and Brooks (1968).
57 **Water power speculators renewed and strengthened:** Graham (1971); Goldberg, R.A. (2003). *America in the Twenties*. New York: Syracuse University Press; Andrews (2006).
57 **The spread of electricity:** Parrish (1992); Brox, J. (2010). *Brilliant: The Evolution of Artificial Light*. New York: Houghton Mifflin Harcourt.
57 **Thanks to the assembly line:** Morison, S.E. (1965). *The Oxford History of the American People*. New York: Oxford University Press.
57 **New products came forth each year:** Parrish (1992).

57 **Fish was promoted as a healthy choice:** e.g., Crie, H.D. (1926). *Fourth Biennial Report of the Commission of Sea and Shore Fisheries of the State of Maine.* Augusta, ME.

58 **Clarence Birdseye:** Kurlansky (2012); also Library of Congress (2010). Who invented frozen food? Everyday Mysteries, loc.gov/rr/scitech/mysteries/frozenfood.html; National Frozen & Refrigerated Foods Association. 2015. History of frozen foods is long and varied, nfraweb.org/resources/articles/details.aspx?ArticleId=18.

58 **Stripped of the associations with unique places:** Mintz, S. (1996). *Tasting Food, Tasting Freedom: Excursions into Eating, Culture, and the Past.* Boston, MA: Beacon Press, p. 114; Gabaccia, D.R. (1998). *We Are What We Eat.* Cambridge, MA: Harvard University Press.

58 **Calvin Coolidge went fishing on the Brule River:** Wisconsin Historical Society (1928). Calvin Coolidge fishing (photograph), http://www.wisconsinhistory.org/Content.aspx?dsNav=N:4294963828-4294955414&dsRecordDetails=R:IM2093. Coolidge was not serious in his announcement, but people took him at his word and he did not seek re-election.

58 **"subtle yet rigid wall of social decorum":** Wilson, J.H. (1975). *Herbert Hoover: The Forgotten Progessive.* Boston: Little, Brown and Company.

59 **an intelligent Irish woman:** Allen, A.B. (2000). *An Independent Woman: The Life of Lou Henry Hoover.* Greenwood Publishing.

59 **When the directors of the Penobscot Salmon Club arrived at the White House:** Hoover, H. (1963). *Fishing for Fun – and To Wash Your Soul.* New York: Random House.

59 **"it is not . . . the function of the government to relieve individuals of their responsibilities":** White, R. (1991). *"It's Your Misfortune and None of My Own": A New History of the American West.* University of Oklahoma Press.

59 **While he continued some major federal construction projects:** Swain (1963); Cahn, R. (1978). *Footprints on the Planet: A Search for an Environmental Ethic.* New York: Universe Books.

60 **eighty potential power sites, including several of "considerable value":** Army Corps of Engineers (1930). Letter from the Secretary of War transmitting report from the chief of engineers on Penobscot River, ME, covering navigation, flood control, power development, and irrigation, House of Representatives Document 652, 71st Congress, Third Session. With experience constructing military fortifications, the Army Corps of Engineers emerged as the main available source of engineering expertise, and so was asked to conduct topographic surveys and assess river and harbor improvements. With the 1824 Rivers and Harbors Act it became Congress's general contractor for constructing water resource development projects, which were believed to have value for both military purposes and interstate commerce; Andrews (2006).

60 **"What are we doing to save the salmon . . .":** Crie, H.D. (1934). *Eighth Biennial Report of the Commissioner of Sea and Shore Fisheries.* Augusta, ME.

61 **he spent hours building pools by hand:** Mares (1999), quoting Dennis, R. (1986). *The Homes of the Hoovers.* West Branch, IA: Herbert Hoover Presidential Library.

61 **the era of executive laxity ended:** Swain (1963).

CHAPTER 6: WASSATAQUOIK

63 **By the 1830s large-scale logging operations were moving up the East Branch:** Waterman, L., and G. Waterman (1989). *Forest and Crag: A History of Hiking, Trail Blazing, and Adventure in the Northeast Mountains.* Boston, MA: Appalachian Mountain Club.

63 **Mr. and Mrs. William Hunt established their homestead:** Foster and Atkins (1868); Kendall (1935).

64 **"stream of light":** Eckstorm translates as "fish-spearing river."
64 **"I do not remember that we saw the mountain":** Thoreau, H.D. (1857). *Journals, from The Writings of Henry David Thoreau*, Journal Manuscript Volume 24, July 31-November 5, 1857. Santa Barbara, CA: University of California, http://thoreau.library.ucsb.edu/writings_journals.html.
64 **investigators with the Maine State Scientific Survey:** Krohn, W.B., ed. (2005). *Manly Hardy (1832-1910): The Life and Writing of a Maine Fur-Buyer, Hunter, and Naturalist*. Northeast Folklore 38. Orono, ME: Maine Folklife Center; Henderson, K.A., ed. ca. (1930). Penobscot East Branch in 1861: From the Diary and Letters of Alpheus Spring Packard. *Appalachia*, pp. 414-426.
64 **The East Branch Penobscot River is ideal spawning habitat:** Cutting (1959).
66 **Burying the eggs in a gravel nest allows more offspring to survive:** Hendry and Stearns (2004); Stearley (1992).
66 **Salmon evolved this complicated life cycle:** Fleming, I.A., and S. Einum (2011). Reproductive ecology: a tale of two sexes, pp 33-65 in Aas et al. (2011); Thorstad et al. (2011).
66 **Overwintering at sea enhances growth:** Hendry and Stearns (2004).
66 **salmon have to enter early:** Hendry, A.P., V. Castric, M.T. Kinnison, and T.P. Quinn (2004). The evolution of philopatry and dispersal: homing versus straying in salmonids, pp. 52-91 in Hendry and Stearns (2004); Thorstad et al. (2011); National Marine Fisheries Service (2012).
66 **settlers fought among themselves:** Pawling (2007).
66 **as part of the regular duties connected with small farms:** Smith (1898).
67 **"Nearly every summer for many years":** Foster and Atkins (1871).
67 **"Running to mountains and shore":** Hindar, K., J.A. Hutchings, O.H. Diserud, and P. Fiske. Stock, recruitment and exploitation, pp. 299-325 in Aas et al. (2011). Netboy (1974) cites another translation, "the salmon shall go from the beach to the mountains, if it wish to go," and dates the law to 900.
67 **the earliest advocates for access to fish were the Penobscots:** Pawling (2007), p. 27. (Recorded) protests occurred in 1801, 1807, 1810, 1812.
67 **Towns could collect penalties from mill owners:** Mill owners also had to allow boats and logs to pass; while the Penobscot was "navigable" only to the head of tide, much of the rest of the river was a "floatable" public highway. See 1919 decision, also *Veazie v. Dwinel*, 50 Me. 479, 484 (1862). Unlike in the West, where the federal government took the lead in altering riverscapes, most Maine rivers were only "navigable," and subject to federal jurisdiction, to the head of tide. The majority of channels were therefore outside of Congressional oversight. While sawmills were fewer in number, pulp and paper mills continued to build dams and reservoirs under the Mill Act, although it was customary to seek legislative sanction for storage dams, Smith (1972), pp. 44-45.
67 **Citizens of Bangor, concerned about the salmon runs:** Godfrey (1882), p. 579.
67 **"The white people take the fish":** Address by Penobscot Chief John Neptune to Maine Governor William King, Maine Senate Chambers, Portland, 11 July 1820, in Godfrey (1882), p. 593.
68 **The lawmen should focus their efforts upriver:** William Wardwell and 175 others (1821). Petition of Penobscot Bay weir fishermen, Maine State Archives Legislative GY, Box 8, File 19.
68 **Several Penobscot fishermen were murdered:** Letter to James Sullivan from Horatio Balch, 7 July 1807, quoted by Pawling (2007).
68 **Joseph Butterfield petitioned the legislature:** Joseph Butterfield petition regarding fisheries on the Penobscot River, Maine. Maine, Executive Council, Indians, 1822 folder, Maine

State Archives, available at http://www.windowsonmaine.org/fullrecord.aspx?objectId=4-113.
68 **Gluskabe went wandering about:** Speck, F. (1918). Penobscot transformer tales. *International Journal of American Linguistics* 1(3):187-244.
69 **Natives and settlers upriver from the commercial fishery:** Stilwell, E.M., and H.O. Stanley (1872). *Sixth Report of the Commissioners of Fisheries of the State of Maine*. Augusta: Sprague, Owen & Nash; see also Baum (1997).
69 **The fishermen saw the mill owners as wealthy:** R. Judd, unpublished notes; Maine State Archives 1836-1855 boxes 134, 127, 132, 18, 158, 170, 155, 192, 164, 174, 214, 222, 195.
69 **hundreds of confusing and contradictory special laws:** Judd (1997), pp. 134-135. Atkins (1887) reported that between 1820 and 1880, the Maine Legislature passed 433 acts related to fisheries, of which 161 pertained to sea-run fish; 71 between 1820 and 1840; 29 in 1840-1860, and 61 in 1860-1880.
70 **The governor appointed:** Foster and Atkins (1869); Elliot, H. (1961). Fish and Game Management in Maine: The Early Years. *Maine Fish and Game Magazine* Spring 1961:13.
70 **"The interest in the preservation and increase of the salmon":** Baird, S.F. (1874). *Report of the Commissioner for 1872 and 1873*, Part II, U.S. Commission of Fish and Fisheries. Washington, DC: Government Printing Office.
70 **"for fish that cannot be caught":** Communication of Benjamin Shaw, Fish Warden, relating to fishing on the Penobscot River, 7 July 1848. Maine State Archives, Legislative GY, Box 195, File 5. Warden Benjamin Shaw even went so far as to praise the dam owners: "The very great bodies of water held back for the purpose of driving logs and the addition of the Allagash have now a tendency greatly to improve the river during the journey of the fish, of which nearly or quite balances the obstructions caused by the Mills and lumbering."
70 **"a residue of backcountry indolence":** Judd (1997).
70 **"wealthy influential few":** Ibid.
71 **"The interposition of artificial dams":** Baird, S.F. (1884). *Report of the Commissioner for 1882*, Part X, U.S. Commission of Fish and Fisheries. Washington, DC: Government Printing Office, p. 57.
71 **The only way forward:** Atkins (1874).
71 **"the art of artificial propagation":** Baird, S. (1880). The Atlantic Salmon, pp. 28-32 in *Report of the Commissioner for 1878*, Part VI, U.S. Commission of Fish and Fisheries. Washington: Government Printing Office. After Baird's death in 1887, fish culture would become the dominant focus of the federal fish commission; Guinan, J.A., and R.E. Curtis (1971). *A Century of Conservation*. National Marine Fisheries Service, http://www.nefsc.noaa.gov/history/stories/century.html.
71 **The state had been experimenting:** Moring, J.R. (1999). Anadromous stocks, pp. 665-696 in *Inland Fisheries Management in North America*, Second Edition (C.C. Kohler and W.A. Hubert, eds.). Bethesda, MD: American Fisheries Society.
71 **asked Atkins to locate a suitable site:** Atkins (1874).
72 **Determined to find a supply:** Ibid.
72 **noting many new redds:** Atkins and Foster (1871).
73 **Breeding was assisted by a "spawn-taker":** Atkins (1874).
73 **Atkins produced 1,560,000 eggs in 1872:** Ibid.; Smith (1897).
73 **The hatchery came under the jurisdiction:** The hatchery was moved briefly to Silver Lake, near Bucksport, but operations returned to Craig Brook in 1879, where they have continued ever since. Anonymous (1874). Collecting salmon spawn in Maine. *Harper's New Monthly Magazine*, June; Atkins, C.G. 1894. Methods employed at Craig Brook Station in Rearing Young Salmonid Fishes, pp. 221-228 in Bulletin of the United States Fish

Commission Volume XIII for 1893. Washington, DC: Government Printing Office; Locke, D.O. (1969). A century of fish culture in Maine. *Maine Fish and Game Magazine*; Baum (1997); Moring, J.R. (2000). The creation of the first public salmon hatchery in the United States. *Fisheries* 25(7): 6-12.

73 **"Making salmon" quickly became an easy solution:** Baird, S. (1876). *Report of the Commissioner for 1873-74 and 1874-5*, Part III, U.S. Commission of Fish and Fisheries. Washington, DC: Government Printing Office.

73 **They think the opportunities for natural reproduction:** Smith, H.M. (1898). Report of the Division of Statistics and Methods of the Fisheries. pp. 125-146 in *Report of the Commissioner for 1897*, Part XXIII, U.S. Commission of Fish and Fisheries. Washington, DC: Government Printing Office.

73 **the U.S. Fish Commission credited the work of Atkins:** Baird, S.F. (1879). *Report of the Commissioner for 1877*, Part V, U.S. Commission of Fish and Fisheries. Washington, DC: Government Printing Office, pp 36-37. See also *Forest and Stream* 32(24), 4 July 1889, p. 485.

73 **"Salmon are no longer a luxury here":** Earll, R.E. (1887). The Coast of Maine and its Fisheries, Part I in Goode, G.B. *The Fisheries and Fishery Industries of the United States*. Section II A Geographical Review of the Fisheries Industries and Fishing Communities for the year 1880, U.S. Commission of Fish and Fisheries. Washington, DC: Government Printing Office.

73 **Recorded landings of salmon peaked in 1888:** Department of Inland Fisheries and Wildlife (1982). *Statewide Fisheries Management Plan*. Augusta, ME.

74 **delivered to New York in a day:** Baird, S.F. (1889). The Sea Fisheries of Eastern North America, Appendix A in *Report of the Commissioner for 1886*, Part XIV, U.S. Commission of Fish and Fisheries. Washington, DC: Government Printing Office.

74 **each Penobscot fisherman had to build an ice-house:** Atkins (1887).

74 **They bought salmon for six cents a pound:** Ibid.

74 **more than one hundred were taken on the river at Hunt Farm:** Stilwell, E.M., and E. Smith (1880). *Report of the Commissioners of Fisheries and Game, of the State of Maine, for the Year 1880*. Augusta, ME: Sprague & Son; Steele, T.S. 1880. The East Branch of the Penobscot, Part Four. *Forest and Stream* 14(1): 3, 5 February 1880.

74 **"as a necessary measure to preserve the remnant of our fish":** Stilwell, E.M., and H.O. Stanley (1873). *Seventh Report of the Commissioners of Fisheries of the State of Maine*. Augusta: Spraque, Owen & Nash. Dynamite could blast twenty-five fish from a single pool.

74 **Stilwell, Maine Commissioner of Fisheries and Game, wrote a letter:** *Forest and Stream*, 17 December 1885.

75 **When he returned to the East Branch:** Smith (1897a).

75 **too few salmon surmounted the dams:** Cutting (1959).

75 **"We have no objections":** Salmon at Veazie. *Bangor Daily Commercial*, 15 July 1916; Veazie Dam fatal bar to salmon. *Bangor Daily News*, 21 July 1916; Salmon netters pay heavy fine. *Bangor Daily News,* 25 July 1916.

75 **Unlike the Great Lakes region:** Jenkins (2003).

76 **The dramatic alterations in flow:** Johnsen, B.O., et al. Hydropower development - ecological effects, pp. 351-385 in Aas et al. (2011).

76 **Bulldozing streambeds in preparation for log driving:** Bond, L.H., and S.E. DeRoche (1950). *A preliminary survey of man-made obstructions and logging practices in relation to certain salmonid fishes of northern Maine*. Augusta: Maine Department of Inland Fisheries and Game. The mainstem log drive ended in 1927-28 (some sources say 1933), although the drive continued on the West Branch until the 1970s.

Notes

76 **heavy rafts of logs tumbling over the rips:** See, for example, description in New England-New York Interagency Committee (1957), Part Two, IX-28.
76 **a three-acre jam:** Smith (1972), p. 91. Another jam, including logs frozen with ice, took out a piece of a Bangor bridge in 1902.
76 **An estimated 4 percent of all logs:** Davies (1972).
76 **effects on the river persist to this day:** National Research Council (2004); Bonney, F. *Maine Sunday Telegram*, 13 August 2006; Fay et al. (2006); P. Larsen, personal communication, 14 August 2014; Kircheis and Liebich (2007).
77 **Frank Perkins reported seeing salmon:** Donahue, J. (1909). *30th Report of the Commissioner of Sea and Shore Fisheries of the State of Maine for 1907 and 1908*. Waterville: Sentinel Publishing Co.; Kendall (1935), Pratt (1946), Cutting (1959). William Converse Kendall, born in Freeport, Maine in 1861, was one of the earliest ichthyologists (fish biologists) in America, and became an international authority on trout and salmon, AuClair (2014).

CHAPTER 7: EMPTY NETS

80 **Such is the story of the first salmon:** Geagan, B. The salmon pool opens today. *Bangor Daily News*, 1 April 1939; Two salmon taken at pool. *Bangor Daily News*, 3 April 1939.
80 **Henrietta Nesbit:** Seale (1986), p. 929. See also Whitcomb and Whitcomb (2000), p. 305.
80 **Roosevelt built a new kitchen for himself:** Seale (1986).
80 **he could peacefully devour kippered herring:** Nesbitt, H. (1951). *The Presidential Cookbook: Feeding the Roosevelts and Their Guests*. Garden City, NY: Doubleday & Co.; Whitcomb and Whitcomb (2000).
80 **Eleanor Roosevelt:** Cannon and Brooks (1968).
80 **Roosevelt went fishing:** Franklin D. Roosevelt Day by Day, http://www.fdrlibrary.marist.edu/daybyday/daylog/august-18th-1939/; Log of the President's cruise onboard the U.S.S. *Potomac* and U.S.S. *Augusta*, 3-16 August 1941, Franklin D. Roosevelt Presidential Library and Museum, Safe Files, Box 1, Folder 07, Atlantic Charter Index, http://docs.fdrlibrary.marist.edu/PSF/BOX1/T07Y06.HTML.
80 **Souvenirs from these and other fishing exploits:** The space became known as "The Fish Room." President Kennedy continued the room's nautical theme by mounting a sailfish that he caught in Acapulco, Mexico. The White House Museum. n.d. The Roosevelt Room, http://www.whitehousemuseum.org/west-wing/roosevelt-room.htm; see also Johnson, C.T. (1970). *A White House Diary*. New York: Holt, Rinehart and Winston, p. 56; Seale (1986), p. 984.
80 **planked and roasted (his favorite):** Seale (1986).
81 **Americans in the 1930s:** Cutler, P. (1985). *The Public Landscape of the New Deal*. New Haven: Yale University Press, p. 146.
81 **The Works Progress Administration:** Ibid., p. 7.
81 **An undercurrent of environmental concern:** Daynes and Sussman (2010), p. 35.
81 **thousands of families were forced to leave their land:** Brox, J. (2010). *Brilliant: The Evolution of Artificial Light*. New York: Houghton Mifflin Harcourt, p. 192.
81 **The 726-foot Hoover Dam:** Lowry, W.R. (2003). *Dam Politics: Restoring America's Rivers*. Washington, DC: Georgetown University Press.
81 **New Deal water development policies:** Andrews (2006), p. 165.
81 **The Federal Power Commission envisioned full development:** Smith, L. (1951). *The Power Policy of Maine*. Berkeley, CA: University of California Press.

82 **But most of Maine's landscape:** Except, perhaps, Roosevelt's plan to place a dam across Passamaquoddy Bay to generate power from the thirty-foot tides, a New Deal proposal that waxed and waned for decades. The working model of the Quoddy Tidal Dam can still be viewed in Eastport, Maine.

82 **"Sawdust as a deterrent is a thing of the past":** Salmon report indicates present crisis, future plenty. *Bangor Daily Commercial*, 3 August 1939.

82 **Atlantic salmon rarely used the Mattaseunk fishway:** In 2011, when 3,090 salmon were documented passing over the Veazie Dam, only 194 were counted going over the Weldon Dam at Mattaceunk, the fifth and final dam across the mainstem Penobscot, USASAC (2012).

82 **salmon could pass most dams:** Holbrook, C. (n.d.) How has fish passage on the Penobscot River changed with the development of dams and fishways since 1820? Unpublished manuscript prepared for NOAA Fisheries, Orono, ME.

83 **Construction of the Grand Coulee Dam:** Eicher, G.J. Fish passage, pp. 163-171 in Benson (1970); Rizzo, B. Fish passage for salmon on the Penobscot, Merrimack, and Connecticut River Basins in New England, in Bohne and Sochasky (1975); see also Clay (1961).

83 **"All I can hope is that the salmon will approve":** Roosevelt to Harold L. Ickes, June 17, 1935, p. 383 in Nixon, E.B., ed. (1957). *Franklin D. Roosevelt and Conservation, 1911-1945*. Hyde Park, NY: General Services Administration, National Archives and Records Service, Franklin D. Roosevelt Library.

83 **he was particularly interested:** Ibid., p. 409.

83 **"The run of salmon is over":** Ickes, H. (1954). *The Secret Diary of Harold L. Ickes, Vol. II, The Inside Struggle 1936-1939*. New York: Simon and Schuster, p. 494. Ickes was among the Republicans who had split from William Howard Taft and the National Party and formed the Progressive Bull Moose Party that re-nominated Theodore Roosevelt in 1912. See also Taylor, N. (2008). *American Made: The Enduring Legacy of the WPA: When FDR Put the Nation to Work*. New York: Bantam Books, p. 83.

83 **"the Bonneville Dam caused a few days delay":** Clay (1961), pp. 11-18.

83 **WPA crews had to retrofit existing dams:** The Bangor Dam fishway was first installed shortly after dam construction in 1874, then rebuilt in 1909 and 1923 and by the WPA in 1936. Neglected, much of it eventually rotted away. Since construction of the Veazie Dam in 1833, fish had been able to pass through a channel at the dam's eastern end. A concrete fishway was added in 1919 and rebuilt by the WPA in 1936, see Cutting (1959).

83 **One experiment found that of sixty salmon:** Pratt (1946).

84 **doomed to spend the summer in the polluted river:** Cutting (1959), and NOAA Fisheries unpublished data. The dams on the main channel of the Penobscot River are "run-of-the-river" dams that slow rather than hold back the river flow, as opposed to storage dams that control water flows. Flashboards provide a means of increasing power generating capacity by raising the elevation of the water, increasing the head or pressure difference.

84 **"It should be possible through careful study":** Salmon report is received by C. of C. *Bangor Daily Commercial*, 6 September 1939, citing Herrington and Rounsefell (1941).

84 **In 1941, a grand total of six adult salmon:** Herrington and Rounsefell (1941).

84 **stocking alone was not enough:** Ibid.

84 **When runs began to decline:** Rounsefell, G.A. (1947). The effect of natural and artificial propagation in maintaining a run of Atlantic salmon in the Penobscot River. *Transactions of the American Fisheries Society* 74:188-208.

85 **a long-range program designed to restore salmon runs:** Rounsefell and Bond (1949).

85 **The current quickened in the ebb tide:** Osborne, O. Salmon pool gives up prize on opening day. *Bangor Daily News*, 2 April 1947, pp. 1, 15.

85 **Local fishing guide Charley Miller:** Bangor guide wants to cook outdoors salmon for President; Season's first salmon shipped to President. *Bangor Daily News*, 3 April 1947, pp. 1, 14.
85 **"There's something to that, Frank!":** photo caption, *Bangor Daily News*, 24 May 1949, p. 1.
85 **Truman was neither picky nor greedy about food:** Truman Trivia Archive, 18 September 1998, Harry S. Truman Library and Museum, http://www.trumanlibrary.org/whistlestop/dailyfastfacts/ff91898.htm.
86 **Truman advocated for conserving resources:** Daynes and Sussman (2010).
86 **he encouraged Americans to avoid meat:** Truman, H.S., Radio and Television Address Concluding a Program by the Citizens Food Committee, 5 October 1947, accessed via The American Presidency Project, http://www.presidency.ucsb.edu/ws/?pid=12766.
86 **per capita consumption of fish and seafood products:** Greenleaf, A.R. (1944). Thirteenth Biennial Report. Department of Sea and Shore Fisheries. Augusta, ME.
86 **Canned fish became regular table fare:** Jarvis (1988), 185. Canned tuna was first packed commercially in 1909; Kurlansky (2012).
86 **luncheon menus featured salmon croquettes:** White House Menu Files, 1947-1952, Bess W. Truman Papers, National Archives, http://research.archives.gov/description/4713727.
86 **tuna America's most popular fish:** Smith, A.F. (2012). *American Tuna.* Berkeley, CA: University of California Press.
86 **The war changed the way Americans ate:** Kurlansky (2012).
86 **Rachel Carson, writing at the time:** Carson, R.L. (1943). Food from the Sea: Fish and Shellfish of New England, *U.S. Fish and Wildlife Service Conservation Bulletin* 33. Washington, DC: Government Printing Office.
86 **No salmon recipes appeared:** Maine Development Commission (1945). The State of Maine's Best Seafood Recipes. Augusta, ME.
86 **salmon that evaded the weirs:** Pratt (1946), pp. 37-53.
86 **Maine's newly created Atlantic Sea-Run Salmon Commission:** Baker et al. (1947).
86 **Atlantic salmon could access only 2 percent:** Cutting (1959); NOAA Fisheries unpublished data.
87 **"There is no doubt that commercial waste":** Baker et al. (1947). While the authors criticized dams without fish passage, the commissioners also recommended water control dams to hold a sufficient volume of water to maintain flow during drought.
87 **In their first report:** Rounsefell and Bond (1949).
88 **There was no limit to the height:** e.g., Eicher, G.J. Fish passage, in Benson (1970), p. 170.
88 **"maximum sustainable yield":** Nielsen, L.A. (1999). History of inland fisheries management in North America, pp. 3-30 in *Inland Fisheries Management in North America* (C.C. Kohler and W.A. Hubert, editors). Bethesda, MD: American Fisheries Society.
88 **Aldo Leopold called for:** Nash (2001), p. 188.
88 **"Flood control dams":** Leopold, A. (1941). Wilderness as a land laboratory. *The Living Wilderness* 6:3.
88 **With diesel engines, bottom trawls:** Bolster (2012).
89 **Flannel-clad Bangor resident Guy Carroll:** *Bangor Daily News*, 1 June 1954, p. 1.
89 **"You know, the only time I've ever caught a salmon:"** As told to 1953 Presidential Salmon angler Walter Dickson. Dickson's fish was the first first fish taken from the shore rather than from a peapod or canoe. Four men remember unforgettable thrill. *Bangor Daily News*, 1 May 1982. See also Mares (1999), p. 16, quoting Mason, E. Ike's first fishing hole. *Rod and Gun Magazine*, March 1955:5, 42.

89 **Broiled fillet of trout was one of his favorite meals:** Rysavy, F. (1957). *White House Chef.* New York, NY: Van Rees Press, p. 78; Smith (1967); Fran Smith Johnson Papers, Dwight D. Eisenhower Library, http://www.eisenhower.archives.gov/research/finding_aids/pdf/Johnson_Fran_Papers.pdf.

89 **Eisenhower needed to prepare for Winston Churchill's upcoming visit:** Secret Service says "no" to president on Maine trip; salmon presented; Churchill coming here for vital talks with President. *Bangor Daily News*, 16 June 1954. Eisenhower's fishing trip to Maine would wait until June 1955, when he fished the Magalloway River near Rangeley with guide Don Cameron. Pankratz, H. (2006). A Guide to Historical Holdings in the Dwight D. Eisenhower Library: Sports and Recreation; http://www.eisenhower.archives.gov/research/subject_guides/pdf/Sports_and_Recreation.pdf.

89 **Domestic issues, such as energy:** Goodwin, C.D. (1981). Energy: 1945-1980. *Wilson Quarterly* 5:55-69. By the 1950s, America had gone from being a net exporter of oil to a net importer.

90 **Eisenhower's First State of the Union address:** Eisenhower, D.D. (1953). Annual message to the Congress on the State of the Union, 2 February 1953. "The best natural resources program for America will not result from exclusive dependence on Federal bureaucracy. It will involve a partnership of the States and local communities, private citizens, and the Federal Government, all working together. This combined effort will advance the development of the great river valleys of our Nation and the power that they can generate. Likewise, such a partnership can be effective in the expansion throughout the Nation of upstream storage; the sound use of public lands; the wise conservation of minerals; and the sustained yield of our forests," http://www.eisenhower.archives.gov/.

90 **The proposed dam had ignited a national debate:** Nash (2001), pp. 209-210. "Not since Hetch Hetchy had so many Americans so thoroughly debated the wisdom of preserving wilderness."

90 **"Atlantic salmon were formerly present":** New England-New York Interagency Committee (1957), p. X-6.

91 **The committee acknowledged that if the complete inventory power plan:** Ibid., pp. V-73, IX-38, IX-44.

91 **only an inventory:** Ibid. After the New England-New York Interagency Committee's inventory work in the 1950s, the region experienced significant floods that required new attention from the interagency committee, whose name was changed to the Northeastern Resources Committee.

91 **adding two new machines at a cost of $38 million:** Sambides, N. Jr., and K. Miller. Oil costs could close Millinocket mill. *Bangor Daily News,* 30 May 2008.

CHAPTER 8: PENOBSCOT

93 **the Mattawamkeag River:** Three dams historically blocked the main channel of the Mattawamkeag, including one at Gordon Falls. Lennon, R.E. (1978). *Mattawamkeag River Studies*. Orono, ME: University of Maine.

95 **changes happen inside the fish:** McCormick et al. (1998).

95 **The transition takes energy:** On average 1.31 percent of eggs laid by a single female will survive to the smolt stage, Thorstad et al. (2011).

96 **A fish that passes through a properly designed downstream bypass:** U.S. Atlantic Salmon Assessment Committee (2011); Nieland, J.L. et al. (2013). Dam impact analysis

model for Atlantic salmon in the Penobscot River, Maine. Northeast Fisheries Science Center Reference Document 13-09. Woods Hole, MA: National Marine Fisheries Service.

96 **biologists and engineers assumed:** Moring, J.R. (1999). Anadromous stocks, pp. 665-696 in *Inland Fisheries Management in North America*, Second Edition (C.C. Kohler and W.A. Hubert, eds.). Bethesda, MD: American Fisheries Society. See also Bley (1987); Fay et al. (2006). Radio telemetry studies by Bangor Hydro in 1991 and 1992 showed that 86 percent of smolts migrating past the West Enfield dam were drawn through the turbines when the water was low.

97 **"responsible" for Maine's Indians:** Proctor, R.W. (1942). *Report on Maine Indians*. Augusta, ME: Legislative Research Committee. The Department of Health and Welfare was given the responsibility for Indian Affairs in 1933 after the Forestry Department had had it for three years. Up until 1929 this responsibility had rested with the Governor of Maine and his Council.

97 **They felt the tribes asked for too much:** Ibid., p. 9.

97 **"The suggestion that a Penobscot Indian Forest be created":** Stevens (1952), p. 10.

98 **Showing defiance, imagination, and courage:** Ghere, D.L. (1984). Assimilation, termination, or tribal rejuvenation: Maine Indian affairs in the 1950s. *Maine Historical Society Quarterly* 239-264; Rodgers, W.H. Jr. (2004). Treatment as tribe, treatment as state: the Penobscot Indians and the Clean Water Act. 55 *Alabama Law Review* 815 (2003-2004); Smith, N.N. 2005. The rebirth of a Nation? A chapter in Penobscot history, pp. 407-423 in *Papers of the Algonquian Conference*, University of Manitoba; Walstad, W.A. (2008). *Maine v. Johnson*: a step in the wrong direction for the tribal sovereignty of the Passamaquoddy tribe and the Penobscot Nation. *American Indian Law Review* 32(2):487-508.

98 **Penobscot Governor Albert Nicola:** Mincher, B. (1952). Nicola says Indians being dictated to. *Bangor Evening Commercial*, 26 June 1952, pp. 1, 2. Native Americans were granted citizenship by the United States in 1924, the State of Maine held back the right to vote from the Wabanaki in federal or state elections until 1957, and did not allow them to elect local representation until the 1960s. The tribes still do not have voting representation in the Legislature.

99 **"stand on their edges":** Thoreau, H.D. Journal, 31 July 1857, Manuscript Volume 23, p. 9, http://thoreau.library.ucsb.edu/writings_journals23.html. Granite in the watershed is limited to areas around Katahdin and Mount Waldo.

99 **Rufus Dwinel successfully sued:** Pike (1967).

99 **enough to block boat traffic:** Atkins (1887).

99 **sawdust particles settled out in the estuary:** Meister, A.L. (1958). The Atlantic salmon, *Salmo salar* Linnaeus, of Cove Brook, Winterport, Maine. M.S. thesis, University of Maine; Haefner, P.A. (1967). Hydrography of the Penobscot River (Maine) Estuary. *Journal of the Fisheries Research Board of Canada* 24(7):1553-1571; Shorey, W.K. (1973). Macrobenthic ecology of a sawdust-bearing substrate in Penobscot River Estuary, Maine. *Journal of the Fisheries Research Board of Canada* 30(4):493-497.

100 **"The extensive deposits have in some instances":** Atkins (1874).

100 **Manufacturing pulp required tremendous volumes of water:** Osborn (1974), p. 38.

100 **"already the extensive pulp mills":** Anonymous (1890). Maine prospects. *Forest and Stream* 34(14):270.

101 **he found the bottom everywhere covered with waste pulp:** Kendall (1935), p. 85.

101 **Maine led the nation:** Wilkins (1932); Maine Pulp and Paper Association (2013). History of Papermaking, http://www.pulpandpaper.org/history.shtml.

101 **a law prohibiting the discharge:** Public Laws of the State of Maine, 1978, Chapter 98, pp. 75-77.

101 **cellulose liquor from the Eastern Manufacturing Company:** Walker, C.H. (1930). *Survey and Report of River and Stream Conditions in the State of Maine.* Augusta, ME.
101 **When oxygen levels drop suddenly:** Noga, E.J. (1996). *Fish Disease Diagnosis and Treatment.* St. Louis, MO: Mosby-Year Book, Inc., pp. 55-59.
101 **"the river appears to be in good condition":** Walker, C.H. (1930). *Survey and Report of River and Stream Conditions in the State of Maine.* Augusta, ME.
102 **requested better cooperation from mill owners:** Feyler, R.E. (1936). *Ninth Biennial Report Department of Sea and Shore Fisheries.* Augusta, ME.
102 **the pollution was a blanket:** Bohne and Sochasky (1975).
102 **dissolved oxygen levels reached zero:** Cutting (1959).
102 **imprinted on polluted water:** A. Meister, interview with the author, 22 August 2014.
102 **"Today's Penobscot, with discolored waters":** Federal Water Pollution Control Administration (1967). Conference: Pollution of the Navigable Waters of the Penobscot River and Upper Penobscot Bay and their tributaries, Proceedings, 20 April 1967, Belfast, ME. U.S. Department of the Interior.
102 **Congress passed the first inland water pollution control bill:** Milazzo (2006).
103 **salmon requested by President John F. Kennedy:** Leavitt, B. Outdoors column. *Bangor Daily News*, 28 June 1961.
103 **Most anglers either gave up or traveled east:** Harkavy, J. Return of Atlantic salmon to Penobscot no "fish story." *Associated Press*, 11 April 1983.
103 **"Restoration of the Atlantic salmon to the Penobscot River depends":** Cutting (1959).

CHAPTER 9: TROUBLED WATERS

105 **Maine's Republican Congressman Clifford McIntyre arrived at the lawn:** A *Bangor Daily News* report from 22 May 1964 states that McIntire gave the fish to McNally, however later stories suggest that Johnson never received it. "A fish caught in the Narraguagus River didn't make it to Lyndon Johnson's table in 1964 beacause Johnson didn't want Maine Republicans catching the resulting headlines. He made a few himself with his refusal," Platt, D. *Bangor Daily News,* 1 May 1981, p. 1, 3. See also Day, J., Some remember when salmon was king. *Bangor Daily News*, 3 April 1967: "LBJ refused to let Republican Clifford McIntire grab any publicity by making the presentation."
106 **president whose favorite dish was chili:** Cannon and Brooks (1968); Whitcomb and Whitcomb (2000), pp. 378-379.
106 **Chef Verdon sliced the salmon:** Verdon, R. (1968). *The White House Chef Cookbook.* Garden City, NY: Doubleday & Company, p. 95, recipe for "Poached Salmon Steak."
106 **whether or not they ever visited it:** Abbey, E. (1968). *Desert Solitaire: A Season in the Wilderness.* New York, NY: Ballantine Books (1971), p. 162.
106 **His view of "environmental" issues:** Daynes and Sussman (2010), p. 57.
107 **President Johnson spoke of uplifting the dignity of the human spirit:** "Special Message to Congress on Natural Beauty," 8 February 1965, Public Papers of the Presidents, Lyndon Baines Johnson, 1965, Book 1.
107 **He listened when his supporters demanded:** Nash (2001); Andrews (2006).
107 **On Capitol Hill:** Milazzo (2006); see also Adler, R.W., J.C. Landman, and D.M. Cameron (1993). *The Clean Water Act 20 Years Later.* Washington, DC: Island Press.
107 **Environmental activists in Maine:** B. Townsend, interview with the author, 19 June 2014.
107 **Muskie called for a national program:** Milazzo (2006).

107 **"if America's waters are troubled":** Committee on Public Works (1964). *Troubled Waters*. Washington, DC: United States Senate, 88th Congress, 1st session.
108 **pollution was a nationwide blight:** Johnson, L.B. Remarks at the signing of the Water Quality Act of 1965, 2 October 1965, available at http://www.presidency.ucsb.edu/ws/?pid=27289.
108 **Pollution had forced the closure:** Council on Environmental Quality (1971). *Second Annual Report*. Washington, DC: Government Printing Office.
108 **Johnson stopped in Portland:** Pollution control bill given unanimous support. *Bangor Daily News*, 29 September 1964, p. 8.
108 **"an open sewer," "a wasteland":** McKeon, E. River cleanup hearings set. *Bangor Daily News*, 12 November 1964, pp. 1, 4; US EPA (1980).
109 **an "attitude of special tolerance":** Osborn (1974).
109 **the river was to assimilate as much pollution as possible:** Maine Water Improvement Commission (1963).
109 **Horace Bond urged the water officials to aim higher.** DEP Water Quality Division (1971). *Interim Water Quality Plan for the Penobscot River Basin*. Augusta, ME.
109 **the state reluctantly reclassified most of the river to Class C:** Osborn (1974); Maine Water Improvement Commission (1963).
110 **only the very strongest of salmon were able to reach spawning beds upriver:** Bernier et al. (1995).
110 **Atlantic salmon on their preliminary list:** U.S. Department of Interior (1964). *Redbook—Rare and Endangered Fish and Wildlife of the United States—Preliminary Draft*. Washington, DC: Bureau of Sport Fisheries and Wildlife. List available at https://askabiologist.asu.edu/first-62-list.
110 **The first official list:** Committee on Rare and Endangered Wildlife Species (1966). *Rare and Endangered Fish and Wildlife of the United States, Bureau of Sport Fisheries and Wildlife Resource Publication 34*. Washington, DC: Government Printing Office. By the time Endangered Species Act was passed in 1973, more than 500 North American species had gone extinct.
110 **"braided into a chain of myths and priorities":** Udall, S.L. (1968). *1976 Agenda for Tomorrow*. New York: Harcourt, Brace & World, Inc., p. 14; Udall, S.L. (1965). *The Quiet Crisis*. New York: Holt, Rinehart and Winston.
110 **intended to develop measures:** Everhart and Cutting (1968).
111 **a mecca for sport fishermen:** Penobscot may become mecca for fishermen. *Bangor Daily News*, 10 January 1967, p. 9.
111 **"If the plan is as good as we expect it to be":** Penobscot model for salmon plan. *Bangor Daily News*, 5 October 1966, p. 29.
111 **Harry Everhart and Richard Cutting envisioned:** Everhart and Cutting (1968). The original 1920 Water Power Act did not include or recognize recreation as a public interest, a provision which was added by amendment in 1935; see Bohne and Sochasky (1975).
111 **a historic Supreme Court decision:** Ewert, S.E. (2001). Evolution of an environmentalist: Senator Frank Church and the Hells Canyon controversy. *Montana* 51(spring): 36–51.
111 **Justice William O. Douglas challenged:** Udall, *Secretary of the Interior v. Federal Power Commission et al.*, No. 463, Supreme Court of the United States, 387 U.S. 428, 87 S. Ct. 1712; 18 L. Ed. 2d 869; 1967 U.S. LEXIS 2772; 1 ERC (BNA) 1069; 1 ELR 20117, June 5 1967.
112 **the Federal Power Commission prepared:** Federal Power Commission (1964). *Penobscot River Basin Maine Planning Status Report* (Revised August 1980). Water Resource Appraisals for Hydroelectric Licensing. New York Regional Office.

112 **License applications were pending or being prepared for the other dams:** "Significant potentials" cited for power projects in Penobscot River. *Bangor Daily News*, 9 October 1964, p. 5.
112 **Americans' per capita consumption of electricity rose 436 percent:** Udall, S.L., C. Conconi, and D. Osterhout (1974). *The Energy Balloon*. New York: McGraw-Hill.
112 **250 "mammoth" power plants would be needed:** Milazzo (2006).
113 **International Minerals and Chemicals:** LCP bought the plant in 1982, they continued operations, but the regulatory climate was changing. In 1986, the plant's continuous release of mercury led the EPA to file an administrative study of hazardous waste handling under the Resource Conservation and Recovery Act, a regulatory program created in 1976 that had recently been amended. EPA officials finally paid a visit to LCP Chemicals in 1991. The violations and spills EPA recorded were just the beginning of a long list that would someday occupy an entire wall of the Maine DEP's file room. In 1994, LCP Chemicals went bankrupt and was bought by HoltraChem Manufacturing Company, LLC, whose president was Bruce Davis, former vice president of Hooker Chemical, the company responsible for Love Canal. http://www.maine.gov/dep/spills/holtrachem/index.html
113 **angler Dick Ruhlin got his fly rod:** R. Ruhlin, interview with the author, 19 February 2010.

CHAPTER 10: KENDUSKEAG

115 **each surge and retreat of the tide:** Bangor is considered the head of tide, but high tides can push beyond Bangor to Eddington Bend and Veazie rapids during monthly high tides and storm surges.
116 **as fast as twenty feet per second:** Aas et al. (2011).
116 **salmon and eels were caught beneath the 150-foot tall cliff:** Godfrey (1882). p. 539.
116 **Mr. Timothy Colby ... caught "a fine fat salmon":** Godfrey (1882), p. 641.
116 **"star on the edge of night":** At the end of "Ktaadn," Thoreau wrote, "There stands the city of Bangor, fifty miles up the Penobscot, at the head of navigation for vessels of the largest class, the principal lumber depot on this continent, with a population of twelve thousand, like a star on the edge of night, still hewing of the forests of which it is built, already overflowing with the luxuries and refinement of Europe, and sending its vessels to Spain, to England, and to the West Indies for its groceries,—and yet only a few axe-men have gone 'up river,' into the howling wilderness which feeds it." Thoreau (1864), p.111.
117 **once stood six mill dams:** Wells (1868).
117 **city planners wanted to entomb the last half mile of Kenduskeag Stream:** Metcalf & Eddy (1946). *Improvement of Kenduskeag Stream*. Bangor Room, Bangor Public Library, Bangor, ME.
117 **they narrowed the stream:** McCord, T. A vote for modern: the story of Bangor's urban renewal. *Bangor Daily News*, 29 December 2008.
117 **conditions were ripe for an intestinal disease epidemic:** Carney, W.J. (1960). *Bangor Department of Health Report on Kenduskeag Stream*. Bangor, ME: Department of Health, reprinted in Bernier et al. (1995).
118 **"Can we afford clean water?":** Adler, R.W., J.C. Landman, and D.M. Cameron (1993). *The Clean Water Act 20 Years Later*. Washington, DC: Island Press.
118 **who believed the environment was in crisis:** Graham (1971), pp. 29-30; Milazzo (2006), p. 140.

118 **a new issue for most Americans:** Rosenbaum, W.A. (1973). *The Politics of Environmental Concern*. New York: Praeger Publishers, p. 7.

118 **interest in the environment had grown:** Osborn (1974), p. 81.

118 **Charles Rabeni began studying the effects:** Rabeni, C.F. (1977). Benthic macroinvertebrate communities of the Penobscot River, Maine, with special reference to their role as water quality indicators, Ph.D. dissertation, University of Maine; Rabeni, C.F., and K.E. Gibbs. (1977). Benthic invertebrates as water quality indicators in the Penobscot River, Maine: completion report to the Maine Department of Environmental Protection and the Land and Water Resources Institute, University of Maine; Davies et al. (1999). The biological monitoring went on to become part of Maine's water classification law, and served as a model for other states.

119 **They also ended the log drive:** Great Northern Paper. Application for License for Big "A" Hydroelectric Project FERC No. 3779, Volume IV, Exhibit E, March 1984.

119 **"a river's assimilative capacity should be used":** Prysunka (1975).

119 **a grandfather clause:** The grandfather exemption expired in 1976; Osborn (1974).

119 **an enormous chemical campaign against the spruce budworm:** Lansky (1992). Maine's Forest Practices Act, the first to regulate forest management including pesticide application, was passed in 1989. These chemicals likely affected Atlantic salmon populations in proximity; for example, later studies showed the potential for nonylphenol and a type of estradiol to interfere with the hormones associated with smolting, reducing salinity tolerance. McCormick, S.D., et al. (2005). Endocrine disruption of parr-smolt transformation and seawater tolerance of Atlantic salmon by 4-nonylphenol and 17β-estradiol. *General and Comparative Endocrinology* 142:280-288.

119 **started demanding more of the mills:** Osborn (1974).

120 **"a water quality success story":** US EPA (1980); Davies et al. (1999). In 1986, the East Branch was upgraded to Class AA and the mainstem was upgraded to Class B in 1990. By 1999 the entire mainstem, except for one mile of the Enfield impoundment, was upgraded to Class B. More than 70 percent of the river is classified as class A or AA.

120 **people came back to the river:** New England River Basins Commission (1981), p. 66; Commercial Whitewater Rafting Study Commission, *Whitewater Rafting Report to the Maine Legislature*, March 1983. Augusta, ME: Office of Legislative Assistants.

120 **A team of agency staff and local representatives:** Hay, E., and B. Bock (1974). Memorandum to Regional Director, Conservation Action/Metro Planning Division, U.S. Department of Interior, 6 August 1974; Bureau of Outdoor Recreation (1974). Penobscot Wild and Scenic River Study Information Brochure. Philadelphia, PA: U.S. Department of Interior.

120 **"We've been in business here seventy-five years":** *Bangor Daily News*, 7 January 1975; Paul McCann, letter to Maurice Arnold, 27 August 1975, Penobscot Paddle & Chowder Society Records, Special Collections, University of Maine Fogler Library.

121 **new fishways were added to each of the dams:** US EPA (1980). The West Branch was excluded from these efforts.

122 **water quality had improved:** US EPA (1980).

122 **With nine new fishways and cleaner water:** Reed, N.P., Assistant Secretary for Fish, Wildlife, and Parks, U.S. Department of the Interior, "Keynote Address," in Bohne and Sochasky (1975).

122 **Spawning adults and inch-long fry were seen in the headwaters:** A. Meister, interview with the author, 22 August 2014.

122 **anglers caught more than thirty:** Haskell, R.N. River management: a one purpose goal or a multi-purpose reality? pp. 150-156 in Bohne and Sochasky (1975).

122 **Fishermen cleared the alder-choked road:** E. Baum, interview with the author, 20 August 2014.

122 **"What a remarkable accomplishment":** Leavitt, B. Outdoors column, *Bangor Daily News*, 26 June 1975.

122 **during a spring flood in 1977:** Some sources say the breach occurred in 1976 or even 1978. However, a *Bangor Daily News* story from March 1977 about the city of Bangor considering funding an engineering firm to determine the feasibility of restoring the Bangor Dam, and the Atlantic Salmon Commission's preference to "see a 20-foot hole punched in the middle of the dam" suggests that the breach had not yet occurred (Strout, J. Bangor's deteriorating dam is seen as threat to salmon restoration, *Bangor Daily News*, 19 March 1977, p. 2). A later article stated "It was in the spring freshet of 1977 that ice, high water and old age teamed up to breach a 15-foot hole" that later widened to seventy-five feet (Harkavy, J. Return of Atlantic salmon to Penobscot no "fish story," *Associated Press*, 11 April 1983). Another article from June 1977 stated "Nature threatens to breach the Bangor Dam. A 30-40 foot section of the old wooden dam structure has been damaged severely," Bond, R.D. Return of the Atlantic salmon, *Maine Audubon News*.

123 **Membership in the Penobscot Salmon Club swelled:** R. Ruhlin, interview with the author, 19 February 2010. See also *Bangor Daily News* reports from 10 June 1978, 11-12 July 1978; and *Maine Sunday Telegram* 12 June 1977.

123 **historic Penobscot Salmon Club:** The National Register of Historic Places listed the Penobscot Salmon Club and adjacent Bangor Salmon Pool on 15 September 1976, recognizing the place as "a part of the Historical and Cultural Heritage of our nation that should be preserved as a living part of our community life and development in order to give a sense of orientation to the American people" (National Register Certificate, Maine Historic Preservation Commission).

123 **seeking cooler and deeper water:** Great Northern Paper Company had shut down their hydroelectric dams in the headwaters while mill workers were on strike. As a result, the water level in the river dropped and the shallow current warmed up quickly.

123 **Hundreds of people . . . came downtown to watch:** It was, as Bud Leavitt reported, "the most talked-about midtown happening since the FBI took to measuring the Brady Gang with machine guns on Central Street." *Bangor Daily News*, 29 June 1978.

124 **prompting litter cleanups and civic pride:** Layton, A.B. Jr. Marsh favors statewide salmon law review. *Bangor Daily News*, 27 July 1978; Sleeper, F. It was catch as catch can, *Sports Illustrated*, 14 August 1981, pp. 64, 67.

124 **the salmon seemed to respond:** It wasn't just Maine rivers that saw returning salmon. Hundreds came back to the Connecticut River, and were sighted as far north as the White River in Vermont, 255 miles upstream from the mouth of the Connecticut River.

124 **Game wardens had to patrol the riverbanks:** Smith, R. Return of the salmon in Maine. *The New York Times*, 12 July 1981.

124 **baked it whole inside a brown paper bag:** C. Robbins, email to the author, 11 August 2013.

124 **The Indian Claims Commission Act awarded:** In 1974, US District Judge George Boldt decreed, on the basis of treaties signed during the 1850s, that tribes in Washington State had the right to harvest one-half of all salmon in Puget Sound.

125 **The Penobscot Nation constitutes:** Patterson, B.H. Jr., and G. Humphreys. Memo to the President, 12 November 1976. Box 4, folder "Passamaquoddy/Penobscot Land Claims" of the Bradley H. Patterson Files at the Gerald R. Ford Presidential Library, Ann Arbor, MI, http://www.fordlibrarymuseum.gov/library/guides/findingaid/pattersonfiles.asp

125 **some 2,800 Indians representing seventy tribes marched:** Rohrer, S. (1978). Indians hit the road over backlash in Washington. *The National Journal* 10(34):1353, 26 August 1978.
125 **The tribes, meanwhile, feared that if Ronald Reagan was elected:** Brodeur (1985).
126 **Butch Wardwell landed a 15.5 pounder:** *Bangor Daily News*, 13 May 1981, pp. 1-2; Wilkinson, L. Bush to get fish marked for Reagan. *Associated Press*, 6 May 1981.

CHAPTER 11: WELCOME TO SALMON CITY

127 **who was recovering from a gunshot wound:** President Reagan's recovery was kept secret from the American people in order to protect Reagan's Superman persona. Ivan Mallet had been told the fish was going to Bush because of "scheduling difficulties," Johnson (1991).
127 **And so they chose to invest their faith:** Mattson, K. (2009). *What the Heck Are You Up To, Mr. President? Jimmy Carter, America's Malaise, and the Speech that Should Have Changed the Country*. New York: Bloomsbury. After his defeat, Carter returned to Georgia to find that two of his prized fly rods had been stolen, including one built by the H.L. Leonard Rod Company, Carter, J. (1988). *An Outdoor Journal: Adventures and Reflections*. New York: Bantam Books.
128 **introduced the White House to nouvelle cuisine:** Haller, H. (1987). *The White House Family Cookbook*. New York: Random House, pp. 327-365.
128 **the Reagans usually went to the small study:** Ibid; Reagan (1990); Colacello, B. Ronnie and Nancy. *Vanity Fair*, August 1998; Whitcomb and Whitcomb (2000), p. 435.
128 **The City of Bangor built the original dam:** Daily water supply 3,789,588 gallons. *Bangor Daily Commercial*, 17 July 1918.
128 **The city continued to draw water from the river:** The city obtained rights to draw drinking water from Floods Pond in nearby Otis. The pond is one of the purest water supplies in the nation and hosts a wild native population of rare blueback trout (Arctic char).
128 **generating about 3 percent:** Office of Energy Resources. *Maine Comprehensive Energy Plan 1976 Edition*. Augusta, ME. Hydropower's share of statewide energy consumption had shrank, from 17 percent in 1954 to 10 percent in 1974. The share would drop to 3.5 percent by 1981. Of the total energy produced in Maine in 1974, hydroelectric plants contributed 28 percent, down from 70 percent in 1950. Industry remained the biggest user, as it had been for much of the twentieth century.
128 **Maine had become dependent on petroleum:** Schurr and Netschert (1973). By 1956, Maine and Vermont were the only states not connected with the gas pipeline network.
128 **public utilities were selling electricity at ever decreasing rates:** *Bangor Daily News*, 31 August 1977, p. 24.
129 **Oil prices skyrocketed:** Goodwin, C.D. (1981). Energy: 1945-1980, From John F. Kennedy to Jimmy Carter. *Wilson Quarterly* 5:70-90.
129 **the federal government encouraged:** Federal programs included Project Independence; Emergency Petroleum Allocation Act (1973); Federal Energy Administration Act (1974); Energy Supply and Environmental Coordination Act (1974); Solar Energy Research, Development, and Demonstration Act (1974); the Energy Policy and Conservation Act (1975); and the Energy Conservation and Production Act (1976). Presidents have promoted expanding domestic energy production ever since; Andrews (2006).
129 **"the energy problem":** The New England Governors' Conference recommended "expeditious implementation of feasible hydroelectric projects." Forecasts for reduced foreign oil consumption in 1985 included a 1150 MW nuclear power plant on Sears Island in Penobscot

Bay, 17 MW from "hydro reclamation" in Maine, and drilling the outer continental shelf/ Georges Bank, New England Regional Commission (1976). A New England Energy Policy, 7 November 1975, pp. I-1 to I-6 in *New England Energy Situation Alternatives for 1985*. Boston, MA: Federal Energy Administration Region I; Office of Energy Resources (1975). *Interim Energy Policy and Comprehensive Energy Plan for the State of Maine, Report to the 107th Legislature*. Augusta, ME.

129 **didn't keep people from looking for new dam sites:** Although plans for major federal dams on the St. John River were first proposed in the 1960s, and Franklin Roosevelt's dream of a tidal energy project was still floating around Passamaquoddy Bay, the New England Regional Commission claimed in 1973 that "the only significant options for meeting New England's expanding energy demands were nuclear energy and petroleum products," Energy Policy Staff, July 1973. *Energy in New England 1973-2000*. Boston, MA: New England Reigonal Commission. Shipman, W.D., and C.E. Veazie (1973). *Energy Policy for the State of Maine*. Brunswick, ME: Public Affairs Research Center, Bowdoin College; New England Committee on Energy, Social Goals, and the Environment (1971). *A New England Perspective on Energy and the Environment*. Boston, MA: New England Natural Resources Center. Maine's Congressional delegation conceded defeat of the massive billion-dollar Army Corps of Engineers dams on the St. John River on May 1, 1981, as the president's salmon was being caught.

129 **The New England Energy Task Force:** Hydroelectric Facilities Workgroup, Energy Resource Development Task Force (1976). *A Report on New England Hydroelectric Development Potential*. New England Federal Regional Council.

129 **The Army Corp of Engineers identified:** Ibid.

129 **Maine's Office of Energy Resources:** Office of Energy Resources. *Maine Comprehensive Energy Plan 1976 Edition*. Augusta, ME.

129 **Robert Haskell told the crowd:** Bohne and Sochasky (1975), p. 82.

130 **the City Council rezoned the land:** *Bangor Daily News*, 14 October 1978, p.1, 3.

130 **more than 350 private hydroelectric projects:** Burke, S.H. (1982). Small scale hydroelectric development and federal environmental law: a guide for the private developer. *Boston College Environmental Affairs Law Review* 9(4):815-861.

130 **list of seventy-eight sites:** New England River Basins Commission (1980). *Potential for Hydropower Development at Existing Dams in New England, Volume I: Physical and Economic Findings and Methodology and Volume IV: State of Maine*. Boston, MA. The New England River Basins Commission evolved from the Northeastern Resources Committee which evolved from the 1950s New England-New York Interagency Committee. With the Water Resources Development Act of 1977 and under pressure from Massachusetts, Congress authorized a study of hydropower expansion in the New England region, at both existing and undeveloped sites. Foster, C.H.W. (1984). *Experiments in Bioregionalism: The New England River Basins Story*. Hanover, NH: University Press of New England.

130 **FERC accepted Massachusetts-based Swift River Company's application:** *Bangor Daily News*, 11 February 1981; 23 February 1982.

130 **President Reagan's old employer:** After leading labor unions in Hollywood, Reagan toured the country as a spokesman for General Electric in the late 1950s, touring plants and factories and being hosted by local executives, a hospitality that included salmon fishing on a yacht off Seattle and deep-sea fishing in the Gulf of Mexico. Reagan, R., and R.G. Hubler (1981). *Where's the Rest of Me?* New York: Karz Publishers.

130 **"There are plenty of rivers in the state":** Murphy, W.T. Dam-power friends, foes to square off Tuesday. *Bangor Daily News*, 13 February 1982, p. 6.

130 **D'Errico had been fishing for striped bass:** R. D'Errico, interview with the author, 3 May 2014.
131 **infusing the local economy:** A survey by the Penobscot Salmon Club found that out-of-state anglers spent an annual total $211,000 to fish the Penobscot. Harkavy, J. Return of Atlantic salmon to Penobscot no "fish story." *Associated Press*, 11 April 1983.
131 **"Salmon City, USA":** Harkavy, J. Salmon run again in reclaimed Maine river. *Associated Press*, 10 April 1983.
131 **Richard Ruhlin explained their opposition:** *Bangor Daily News*, 16 May 1981, p. 33.
131 **"rickety, century-old jumble of crumbled cement":** *Bangor Daily News*, 23 February 1982, p. 10.
131 **called the plan to rebuild the Bangor Dam "insane":** Rawson, D. Bangor Dam issue draws 250 people. *Bangor Daily News*, 17 February 1982, p. 1.
131 **"It's asinine and crazy to dam up the best fishing river":** Collins, D. On the Penobscot, the salmon generate the power. *The Washington Post*, 28 September 1982, p. D2.
132 **Before 1980, the state gave priority:** Waite, G.G. (1964). Public rights in Maine waters. *Maine Law Review* 17: 161-204.
132 **"The state should clearly identify river stretches":** Hydropower Study Subcommittee (1981). *Recommended State Policies for Hydropower Development in Maine*. Augusta, ME: Maine Land and Water Resources Council.
132 **incorporated other values of rivers:** New England River Basins Commission (1980). *Potential for Hydropower Development at Existing Dams in New England, Volume I: Physical and Economic Findings and Methodology and Volume IV: State of Maine*. Boston, MA. The commission concluded: "Hydropower development is likely to be a volatile issue in the Penobscot Basin for the next several years. Basinwide, development potential at both existing dams and undeveloped sites is among the highest in New England." When Maine submitted their Comprehensive Hydropower Plan to FERC in 1982, the state acknowledged that "Not all available hydroelectric resources will be developed, because of the desire to recognize valid environmental concerns."
132 **The commissioners rushed to publish their final report:** New England River Basins Commission (1981). *Water, Watts, and Wilds: Hydropower and Competing Uses in New England*. Boston, MA.
132 **They created maps of competing uses:** The maps don't distinguish species and are at a scale of 1:500,000. Existing anadromous fisheries were mapped up the lower Penobscot River, South Marsh Stream, Piscataquis, Pleasant, Seboeis, Mattawamkeag, and East Branch. Mapped as potential fisheries with inaccessible habitat were Sedgeunkedunk, Pushaw/Stillwater, Olamon, Passadumkeag, Big Wilson/Sebec. No anadromous fisheries, existing or potential, were mapped on the West Branch.
133 **"What began with the Bangor Dam rippled statewide":** Platt, D. Dam project must not diminish uniqueness of river. *Bangor Daily News*, 31 August 1983, p. 17; see also Brown, M.T.P. (1984). Environmental watch: a review of victories and defeats in the year just past. *Down East* 30(6):7-12.
133 **Utilities scoffed at these new additional restrictions:** Maine Land and Water Resources Council (1985). *The Maine Rivers Policy, 1983-1985: A Progress Report to the Governor and Legislature*. Augusta, ME: State Planning Office.
133 **When Ronald Reagan was a young boy:** Reagan, R., and R.G. Hubler (1981). *Where's the Rest of Me?* New York: Karz Publishers; Reagan (1990); Johnson (1991).
134 **"handiwork of God":** Reagan (1990). The section on Ronald Reagan's actions and environmental policies as president is compiled from the following sources: Boaz, D., ed. (1988). *Assessing the Reagan Years*. Washington, DC: Cato Institute; Colacello, B. Ronnie

and Nancy. *Vanity Fair*, August 1998; Lash, J., K. Gillman, and D. Sheridan (1984). *A Season of Spoils: The Story of the Reagan Administration's Attack on the Environment.* New York: Pantheon; Johnson (1991); Morris, E. (1999). *Dutch: A Memoir of Ronald Reagan.* New York: Random House; Reagan (1990); Shanley, R.A. (1992). *Presidential Influence and Environmental Policy.* Westport, CT: Greenwood Press; Shabecoff (1993); Hirt, P.W. (1994). *A Conspiracy of Optimism: Management of the National Forests since World War Two.* Lincoln, NE: University of Nebraska Press; Skinner, K.K., A. Anderson, and M. Anderson, eds. (2003). *Reagan: A Life in Letters.* New York: Free Press.

134 **Clean Water Act:** Maine Senator George Mitchell led the effort to keep it. "We could not go back to the days when eighty-five percent of our waterways were polluted." Sen. G. J. Mitchell, interview with the author, 29 September 2014.

134 **a policy of "planned neglect":** This policy was confirmed during questioning of Reagan officials by Maine Senator George Mitchell at the Senate committee meeting on shrinking the Superfund program. Lash, J., K. Gillman, and D. Sheridan. (1984). *A Season of Spoils: The Story of the Reagan Administration's Attack on the Environment.* New York: Pantheon.

135 **Critical habitat for endangered species went undesignated:** Darin, T.F. (2000). Designating critical habitat under the Endangered Species Act: habitat protection versus agency discretion. 24 *The Harvard Environmental Law Review* 209.

135 **Reagan proposed to withhold:** Williams, T. (1981). The long journey home of the the leaper, "king" of the fish. *Smithsonian* 12(November):170.

135 **Public support for environmental protection increased:** Dunlap, R.E. (1987). Polls, pollution, and politics revisited: public opinion on the environment in the Reagan Era. *Environment* 29(6):6-11, 32-37; Andrews (2006); Sen. G. J. Mitchell, interview with the author, 29 September 2014.

135 **Bumper stickers multiplied:** Bud Leavitt Papers, Special Collections, University of Maine Fogler Library.

136 **Great Northern spent $122 million:** Sambides, N. Jr., and K. Miller. Oil costs could close Millinocket mill, *Bangor Daily News*, 30 May 2008, pp. 1, 8.

136 **supplied only about one-quarter:** Federal Power Commission (1964). Penobscot River Basin Maine, Planning Status Report, Water Resource Appraisals for Hydroelectric Licensing; Officials anticipate more hydro power for state. *Bangor Daily News*, 7 August 1984. Six plants on the West Branch develop 413 feet of the total fall of seven hundred feet between Ripogenus and Medway. The seven plants on the mainstem of the river developed 142 feet of the total fall of 240 feet to head of tide. Dams owned by Maine industries provide about 27 percent of total industrial electricity requirements.

136 **Great Northern asserted:** Butterfield, F. Plan for a dam stirs fears for Maine's Penobscot. *The New York Times*, 14 October 1984, p. 26.

137 **came to the river to fish for landlocked salmon:** As in the New England River Basins Commission Report, in the Maine Rivers Study, the West Branch was notable for its geologic features and rare plants of Ripogenus Gorge, its undeveloped and scenic corridors and stretches of whitewater. The authors acknowledged that the West Branch historically contained anadromous fish, but restoration was deemed unrealistic and was "de-emphasized due to potential habitat conflicts with an excellent inland fishery" for stocked, land-locked salmon, Maine Department of Conservation and U.S. Department of the Interior (1982). In the New England River Basins Commission report, the West Branch appeared on maps of scenic whitewater reaches with recreational value, but was not included on the map of sea-run fisheries, not even as "potential with inaccessible habitat."

137 **Outdoor recreation groups like the Appalachian Mountain Club:** The Penobscot Coalition to Save the West Branch included Natural Resources Council of Maine, the Appala-

chian Mountain Club, Maine Audubon Society, Sportsman's Alliance of Maine, Trout Unlimited, the Sierra Club, National Wildlife Federation, the Penobscot Paddle and Chowder Society, New England Fly Tyers, and Moosehead Lake Wilderness Association. The coalition also involved the Eastern Professional River Outfitters Association, and the Boston-based New England Rivers Center.

137 **In courageous testimony:** S. Neily, email to the author, 26 November 2014.

137 **a four-day public hearing on the Bangor Dam:** Tremble, T.J. Board opens 2-day hearing on Bangor dam. *Bangor Daily News,* 18 July 1985, pp. 1, 4.

138 **The salmon fishermen did not:** Platt, D. Swift River seeks dam permits. *Bangor Daily News,* 30 June 1984, p. 21.

138 **Great Northern president Robert Bartlett:** Paper company drops plans for dam in Maine, *Associated Press,* 14 March 1986; Larmer, B. Victory for Maine's wilderness, setback for paper company. *The Christian Science Monitor,* 27 March 1986, p. 3: University of Maine professor and lumber industry historian David C. Smith said that the Big A defeat reflected a weakening of Great Northern's relationship with the state of Maine. "In the last 15 years, since their merger with Nekoosa Redwoods, many people have felt that they didn't have as much of the interests in the state of Maine at heart. . . . There's been a shift away from the symbiotic relationship between Great Northern and the state of Maine. And the Big A defeat must fit into that." Great Northern Nekoosa already owned mills in Panama and Costa Rica before a hostile takeover by multinational Georgia Pacific in 1990. Lansky (1992), pp. 39-40.

138 **In passing the Electric Consumers Protection Act of 1986:** Klein, C.A. (1999). Dams and democracy. 78 *Oregon Law Review* 641, citing H.R. Rep. No. 99-507 (1986). The report observed that "in developing the legislation, the Committee has been keenly aware of the variety of these [public] interests and has fashioned a bill that protects those interests," citing concern for factors beyond the need for power, such as navigation, flood control, irrigation, fish and wildlife protection, recreation, aesthetic considerations, water quality control, impacts on drinking water, and dam safety and adequacy. Maine's Senator Mitchell had tried to introduce legislation to require FERC to comply with state decisions on hydropower planning.

139 **ended the era of big dam projects:** Reisner, M. (1986). *Cadillac Desert.* New York, NY: Viking; McCully, P. (2001). *Silenced Rivers: The Ecology and Politics of Large Dams.* New York: Zed Books.

139 **the eyes of the nation were on Maine:** Ryddle, L. Maine dams removing obstructions to salmon. *The New York Times,* 7 August 1988, p. 39.

139 **Reagan was swapping fish stories:** 87-year-old takes salmon to Reagan, *Bangor Daily News,* 8 May 1987, pp. 1-2.

139 **serving Columbia River salmon at a state dinner:** Egan, T. (1990). *The Good Rain.* New York, NY: Vintage Books; Colacello, B. Ronnie and Nancy. *Vanity Fair,* August 1998. "It was the most significant and highly anticipated of the many state dinners Mrs. Reagan had arranged, and required careful planning. Mrs. Reagan asked renowned pianist Van Cliburn to perform. In 1958, Cliburn had won the Tchaikovsky Competition in Moscow, the first American to have done so. His performance of a Rachmaninoff piece and a rendition of 'Moscow Nights' was received with great enthusiasm from the Soviet delegation, especially Secretary Gorbachev. Mrs. Reagan would later say that the dinner was one of the great evenings of her husband's presidency." Throughout history, cooks and chefs have served salmon to distinguished foreign visitors. Salmon from the Rhine River was shipped to Paris, where Kaiser Wilhelm and Prince Bismarck served it at banquets, Netboy (1974).

139 **he repeated the chorus:** Reagan, R. Remarks on Signing the Annual Report of the Council on Environmental Quality, 11 July 1984, Arlington, VA, http://www.reagan.utexas.edu/archives/speeches/1984/71184a.htm; Reagan, R. Remarks to the National Campers and Hikers Association, 12 July 1984, Bowling Green, KY, http://www.reagan.utexas.edu/archives/speeches/1984/71284a.htm.

140 **Charles Caron of Brewer got to send:** Mainer declines to give his salmon to Reagan, *Associated Press*, 4 May 1988; Baum (1997).

CHAPTER 12: ALAMOOSOOK

142 **Charles Atkins bought hundreds of salmon:** Baird, S. (1882). *Report of the Commissioner for 1879*, Part VII, U.S. Commission of Fish and Fisheries. Washington: Government Printing Office; Steenstra, E.P. (2012). Charles Grandison Atkins. *Eddies: Reflections on Fisheries Conservation* 5(1):8-9.

143 **"Where the salmon go":** Cheney, A.N. (1892). Salmon at sea. *Forest and Stream*, 38(7):152, 18 February 1892. See also Smith, H.M. (1895). Notes on the capture of Atlantic salmon at sea and in the coast waters of the Eastern states, pp. 95-99 in *Bulletin of the United States Fish Commission*, Volume XIV, for 1894. Washington, DC: Government Printing Office.

144 **the occasional appearance of salmon:** Smith (1898).

144 **repeating the same refrain:** Kendall (1935); Maine Atlantic Sea-Run Salmon Commission (1954). *Biennial Report*. Augusta, ME; Dymond (1963); Collette, B.B., and G. Klein-MacPhee (2002). *Bigelow and Schroeder's Fishes of the Gulf of Maine*, Third Edition. Washington, DC: Smithsonian Intitution Press.

144 **a tagged kelt released in Scotland's Blackwater River:** Menzies, W.J.M., and W.M. Shearer (1957). Long-distance migration of salmon. *Nature* 179:790.

145 **the U.S. Navy began sending submarines:** Mowat, F. (1984). *Sea of Slaughter*. Boston: Atlantic Monthly Press. This account does not appear to be repeated in any other reference.

145 **Denmark invested large sums of money:** Netboy (1974); Mills (1989); A. Meister, interview with the author, 22 August 2014.

145 **Construction of modern fish-processing plants:** Paloheimo, J.E., and P.F. Elson (1974). Effects of the Greenland fishery for Atlantic salmon on Canadian stocks, International Atlantic Salmon Foundation Special Publication series 5(1). St. Andrews. Much of the fish was shipped frozen, and exported to Scandinavian countries for smoking.

145 **Catches rose quickly:** Mills (1989); Jensen, J.M. (1990). Atlantic salmon at Greenland. *Fisheries Research* 10:29-52; Reddin, D.G., and K.D. Friedland (1999). A history of identification to continent of origin of Atlantic salmon (*Salmo salar* L.) at west Greenland, 1969–1997. *Fisheries Research* 43(1): 221-235.

146 **Many feared that the sudden multinational, complex fishery:** Rideout, S.G., and J.A. Ritter (2002). Canadian and US Atlantic salmon institutions and politics: where are the fish and who cares, pp. 93-116 in *Sustaining North American Salmon: Perspectives Across Regions and Disciplines*. Bethesda, MD: American Fisheries Society; Whoriskey (2009). Reagan also signed the Pacific Salmon Treaty to limit commercial fishing and jointly manage stocks on the West Coast in 1985 (although talks over acid rain broke down at the same summit with Canadian Prime Minister Brian Mulroney), Sanger, C. North American summit shirks acid rain issue. *Guardian Weekly*, 24 March 1985, p. 6.

146 **the government decided to favor the recreational fishery:** Whoriskey (2009). Commercial fisheries were closed in Newfoundland (1992), Labrador (1998), and Quebec (2000).

146 **at the urging of President Reagan:** Reagan, R. Message to the Senate Transmitting the Convention for the Conservation of Salmon in the North Atlantic Ocean, 22 June 1982, http://www.reagan.utexas.edu/archives/speeches/1982/62282a.htm.

146 **focused at first on debating and developing regulatory measures:** Sochasky, L. (1983). *New England Salmon Management: Proceedings of the Conference on the Management of Atlantic Salmon in New England*, West Springfield, Massachusetts, May 13-14, 1983. Special Publication Series Number 12. St. Andrews, NB, Canada: Atlantic Salmon Federation; Colligan, M., T. Sheehan, J. Pruden, and J. Kocik. 2008. The challenges posed by international management of Atlantic salmon: balancing commercial, recreational, and societal interests—the North Atlantic Salmon Conservation Organization. *American Fisheries Society Symposium* 62:1-22.

146 **NASCO meetings were politically charged and tense:** Butler and Taylor (1992); B. Townsend, interview with the author, 19 June 2014; T. Sheehan, email to the author, 11 November 2014.

147 **instrumental in international negotiations:** Miller, A.S., T.F. Sheehan, R.C. Spencer, M.D. Renkawitz, K.D. Friedland, and A.L. Meister (2012). Description of the historic US Atlantic salmon (*Salmo salar* L.) tagging programs and subsequent databases. Northeast Fisheries Science Center Reference Document 12-13. Woods Hole, MA: National Marine Fisheries Service, http://nefsc.noaa.gov/publications/; Miller, A.S., T.F. Sheehan, M.D. Renkawitz, A.L. Meister, and T.J. Miller. 2012. Revisiting the marine migration of US Atlantic salmon using historical Carlin tag data. *ICES Journal of Marine Science* 69:1609-1615.

147 **When Native Americans first began ceding their lands:** Chavaree, M.A. (1998). Tribal sovereignty. *Wabanaki Legal News* Winter 1998. Pine Tree Legal Assistance.

147 **Sea-run and marine fish and shellfish made up:** Harper, B., and D. Ranco (2009). *Wabanaki Traditional Cultural Lifeways Exposure Scenario*. Boston, MA: Environmental Protection Agency.

147 **the Penobscots netted two salmon:** Leavitt, B. Outdoors column, *Bangor Daily News*, 9 December 1987, p. 15; Tierney, J.T., letter to William Vail, Atlantic Sea-Run Salmon Commission chair, 16 February 1988, Augusta, ME: Department of the Attorney General; J. Banks, interview with the author, 19 August 2014.

147 **they decided not to take any more salmon:** Saucier, R.M. Absent groups limit dialogue at Penobscot River conference; industry, others not represented at stakeholders' gathering. *Bangor Daily News*, 24 June 1996.

148 **Preventing pollution, an act of tribal self-preservation:** US EPA (1991). EPA, Federal, Tribal and State Roles in the Protection and Regulation of Reservation Environments. Washington, DC.

148 **The Penobscots began monitoring:** Cohen, E.B. (2000). Effect of Maine Indian Claims Settlement Act on State of Maine's application to administer National Pollutant Discharge Elimination System (NPDES) Program, Opinion of the Department of the Interior Office of the Solicitor, Washington, DC; D. Kusnierz, interview with the author, 17 October 2014.

148 **Dioxin was one of their most serious concerns:** Miller, D.S. (1994). *Final Report, Analysis of Cancer Registry Data, Penobscot Indian Nation, 1959-1993*. Atlanta, GA: Centers for Disease Control and Prevention; Division of Community Health Investigations (2014). *Public Health Assessment, Review of Sediment and Biota Samples, Penobscot River, Penobscot Indian Nation, Maine*. Atlanta, GA: Agency for Toxic Substances and Disease Registry.

148 **dioxin and mercury contamination:** Williams, R.L., and L. Cseh (2007). A review of dioxins/furans and methyl mercury in fish from the Penobscot River, located near Lincoln, Maine. *Toxicology and Industrial Health* 23(3):147-153.

148 **The tribe's monitoring documented:** The Penobscots participated in negotiations that resulted in EPA issuing "the most stringent dioxin limits ever imposed on a kraft paper mill in the United States," and eventually forced Lincoln Pulp & Paper to stop producing TCDD in 1999. The mill spent more than $100 million to rebuild their bleach plant to eliminate the use of chlorine, Ranco, D. (2014). The trust responsibility and limited sovereignty: what can environmental justice groups learn from Indian Nations? *Society & Natural Resources: An International Journal* 21(4)354-362.

149 **Their data also helped to upgrade classification:** Penobscot Nation Water Resources Program (2002). *Summary of Accomplishments 2001-2002*. Indian Island, ME: Department of Natural Resources; Penobscot Nation Water Resources Program (2005). *General Summary of Water Quality Monitoring Conducted on Penobscot Reservation and Trust Lands*. Indian Island, ME: Department of Natural Resources; Penobscot Nation Water Resources Program (2007). *Water Quality Assessment Report*. Indian Island, ME: Department of Natural Resources; D. Kusnierz, interview with the author, 17 October 2014.

149 **"The last thing the state wanted to see":** Bayly, J., unpublished manuscript excerpt, distributed at Penobscot Indian Nation NPDES Appeal Press Conference, 15 March 2004, Indian Island, ME.

149 **Salmonid stocks were declining worldwide:** Nehlsen, W., J.E. Williams, and J.A. Lichatowich (1991). Pacific salmon at the crossroads: stocks at risk from California, Oregon, Idaho, and Washington. *Fisheries* 16(2):4-21.

149 **the Atlantic Salmon Federation promoted the newly emerging salmon farming industry:** Whoriskey (2009).

149 **Atlantic salmon aquaculture:** Mills (1989); Baum (1997); O'Hara, F., C. Lawton, and M. York (2003). Economic impact of aquaculture in Maine. Hallowell, ME: Planning Decisions, Inc.; Whoriskey (2009).

150 **a mix of strains:** Baum, E. (1998). History and description of the Atlantic salmon aquaculture industry of Maine, Canadian Stock Assessment Secretariat Research Document 98/152, Department of Fisheries and Oceans. Maine salmon farms are now prohibited from using salmon stocks with European ancestry.

150 **cultured fish were documented:** Fish and Wildlife Service and National Marine Fisheries Service, Final Endangered Status for a Distinct Population Segment of Anadromous Atlantic Salmon (*Salmo salar*). *Federal Register*, 17 November 2000, 65 FR 69459.

150 **a massive decrease in the numbers of escaped farm salmon:** Whoriskey (2009).

150 **When Don Shields caught a fish in 1992:** Butler and Taylor (1992).

CHAPTER 13: PAPER SALMON

151 **Salmon Billy:** Presidential Catch (photo caption). *Bangor Daily News*, 2 May 1989; R. Spencer, personal communication, 8 May 2014; B. Townsend, interview with the author, 19 June 2014; E. Baum, interview with the author, 20 August 2014.

152 **Bush arrived in Maine:** Bush goes fishing in Maine. *Bangor Daily News*, 19 May 1989, p. 8.

152 **"slippery ledge") islands:** Eckstorm notes this as "the alewife fishing place below the outlet."

152 **the largest sawmill in the world under one roof:** Smith (1972); Hale (2005).

152 **six million feet of logs:** Rogers, H.W. (1926). *A History of Orono Maine*, bound typescript copy, Special Collections, University of Maine Fogler Library.

152 **water wheels groaning and saws screeching:** Day, C.A. (1956). *Historical Sketch of the Town of Orono*. Special Collections, University of Maine Fogler Library.

152 **the best salmon fishing was often below the Basin Mills:** Foster and Atkins (1871).

152 **sawmills shared the basin:** Peterson, S.D. (2004). *Images of America: Orono*. Portsmouth, NH: Arcadia Publishing.

153 **the probability of restoring salmon:** Astbury, C. Basin Mills: wise use of water power or threat to salmon? *Bangor Daily News,* 5-6 December 1992.

153 **"The public is unaware":** Meeting Minutes, Bud Leavitt Papers, Box 2024, Folder 11, Special Collections, University of Maine Fogler Library.

154 **Basin Mills would interfere with their rights:** Kesseli, D. Effort under way to avoid litigation over dam projects Basin Mills licensing to be delayed. *Bangor Daily News*, 23 February 1996.

154 **"Our position is simple":** Bisulca, P. (1996). Penobscot: A People and Their River. *Conservation Matters* (Summer 1996), available at http://www.penobscotnation.org/Arts/Articles/bisulca.htm.

155 **Department of Interior consistently supported:** Juni, R., and J. Rabinowitz, 2 May 1992 amicus curiae from U.S. Department of Justice; Austin, P. Fishing rights threatened: Justice Department supports Penobscots against Basin Mills. *Maine Times,* 3 June 1994. In approving Basin Mills, the Board of Environmental Protection provided that a few token salmon be released above the dam, so that the tribe could engage in "angling," a practice not in accord with the tribe's understanding of fishing rights. In 1997, Maine Attorney General Andrew Ketterer stated that the Penobscots were never guaranteed a particular quantity of fish. "At the time of the Settlement Acts in 1980, the type of 'sustenance' fishing the Nation now lays claim to did not exist. Indeed, they do not and cannot suggest that the fishery in 1980—or for any appreciable time prior thereto—was sufficient to allow the members of the Penobscot Tribe to live off their catch . . . obviously the fish were fewer and more contaminated." The state believed that the Settlement Acts "wiped the slate clean" with regard to the Penobscots place in Maine, and so they seemed to think that because the Penobscots had not been able to live off salmon and other fish in the river for decades if not centuries, they couldn't expect to in the future. Bayly, J., unpublished manuscript excerpt, distributed at Penobscot Indian Nation NPDES Appeal Press Conference, 15 March 2004, Indian Island, ME, p. 11.

155 **both sides realized they had something to gain:** B. Townsend, interview with the author, 19 June 2014.

155 **Folders of fact sheets and glossy brochures:** Basin Mills folder, Maine Water Resources Research Institute files, now in the collection of the author.

155 **Bangor Hydro liked to show graphics:** Bangor Hydro-Electric Co. Energy for the Future (video), Unity College Library, Unity, ME.

156 **they would abandon their plans:** "Whether the project in the long term can ever be built is a big question, given the changes in the industry," said company spokesman William Cohen. Porter, N. FERC draft no news to opponents of dam; Coalition warns of impact on salmon. *Bangor Daily News*, 15 November 1994.

156 **who believed that being in control of rivers:** Letters to the editor, *Bangor Daily News*, 22 November 1994.

157 **a thing on paper:** Hennessy, T. *Bangor Daily News*, 16 April 1994; AuClair (2014).

158 **upheld the state's approval of the dam:** Basin Mills decision. *Bangor Daily News*, 17 September 1994.

158 **most endangered rivers:** Day, J.S. Environmentalists calling Penobscot endangered river. *Bangor Daily News*, 19 April 1995.

158 **Bangor Hydro would have to sell:** Kesseli, D. Effort under way to avoid litigation over dam projects Basin Mills licensing to be delayed. *Bangor Daily News*, 23 February 1996; Bodo, P. In Maine, signs of hope for rivers and anglers. *The New York Times,* 16 November 1997.

158 **FERC denied Bangor Hydro a license:** Federal Energy Regulatory Commission, Order on Applications for New and Original Licenses, Bangor Hydro-Electric Company, Project No. 10981-000 et al., 20 April 1998, 83 FERC 61,039. The language of FERC's denial made clear that their decision was affected by the Electric Consumers Protection Act of 1986, which mandated "equal consideration" be given to various factors when making a licensing decision, Lowry, W.R. (2003). *Dam Politics: Restoring America's Rivers*. Washington, DC: Georgetown University Press.

159 **a resolution opposing the listing:** Jenkins (2003).

159 **the listing was debated:** Clancy, M.A. Salmon proposal fight urged. *Bangor Daily News*, 7 January 2000.

159 **"a Draconian proposal":** Clancy, M.A. Officials take heat over salmon listing. *Bangor Daily News*, 31 January 2000, pp. A1-A2.

159 **salmon were nothing but "mongrels":** "Enviros" play the salmon card. *Bangor Daily News*, 14 January 2000, p. A11.

159 **If salmon are not surviving:** French, E. Extinction acceptable in Maine? *Quoddy Tides*, 14 January 2000, p. 4.

160 **sued to prevent the listing:** *Maine v. Norton*, 257 F. Supp. 2d 357 - Dist. Court, D. Maine 2003. The claims of the businesses were dismissed by the court.

160 **"heavy-handed regulatory burden":** Bradbury, D. Salmon's high price. *Maine Sunday Telegram*, 27 February 2000, pp. 1A; 4A-5A.

CHAPTER 14: WASSUMKEAG

161 **people knew the island as Brigadier's Island:** Eastman, J.W. (1976). *A History of Sears Island*. Searsport, ME: Searsport Historical Society.

161 **Commissioner Hugh Smith reported:** Smith (1898).

161 **"a monopoly on the qualities by which people judge fish":** Williams, T. (1981). The long journey home of the the leaper, "king" of the fish. *Smithsonian* 12 (November):170.

162 **eels journey to the Sargasso Sea:** Eels go to the Sargasso Sea to spawn, although scientists have not documented actual reproduction. The saltwater-freshwater-saltwater cycle makes them "catadromous." Fish that hatch in freshwater, go to sea, and return to freshwater to spawn are "anadromous"; the collective term for both types of sea-run species is "diadromous."

162 **around 1 percent of their historic numbers:** Fay et al. (2006).

162 **helping to "buffer" salmon smolts:** Schulze, M.B. (1996). Using a field survey to assess potential temporal and spatial overlap between piscivores and their prey, and a bioenergetics model to examine potential consumption of prey, especially juvenile anadromous fish, in the Connecticut River estuary. M.S. thesis. University of Massachusetts, Amherst, MA; Mather, M.E. (1998). The role of context-specific predation in understanding patterns exhibited by anadromous salmon. *Canadian Journal of Fisheries and Aquatic Sciences* 55(S1):232-246; USASAC (2004); Saunders et al. (2006).

163 **Other species act as habitat engineers:** Hogg, R.S., S.M. Coghlan Jr., J. Zydlewski, and K.S. Simon (2014). Anadromous sea lampreys (*Petromyzon marinus*) are ecosystem engineers in a spawning tributary. *Freshwater Biology* 59:1294-1307.

163 **responsible for the towering height of King Pine:** W. Leavenworth, email to Diadromous Species Restoration Research Network, 27 May 2014: "A pine large enough for a one-piece lower main mast for a King's ship must have held 10,000 board feet and been 250 feet tall—as large as an old-growth sitka spruce on the Pacific Coast, where links between tree growth and the import of marine nutrients from sea-run fish are well-established. Today, a 2,000-board-foot tree is unusually large."

163 **"You name it, we caught it":** Taylor, D. (1974). Interview with Avery Bowden, 5 March 1974, MF 049 Penobscot River Commercial Fisheries Project 0807, Maine Folklife Center Collections, University of Maine.

163 **more species showed up in the nets and weirs:** Evermann, B.W. (1905). Salmon Fishery of Penobscot River and Bay, pp. 110-114 in *Report of the Division of Statistics and Methods of the Fisheries, Report of the Commissioner for the Year Ending June 30, 1903*, Part XXIX, U.S. Commission of Fish and Fisheries. Washington, DC: Government Printing Office.

163 **Cod spawned in the channel:** Ames, E.P. (2004). Atlantic cod stock structure in the Gulf of Maine. *Fisheries* 29(1):10-28.

163 **"the chief agency in the decrease":** Baird, S. (1880). The Atlantic Salmon, pp. 28-32 in *Report of the Commissioner for 1878*, Part VI, U.S. Commission of Fish and Fisheries. Washington, DC: Government Printing Office.

164 **the Edwards Dam removal targeted:** FERC relied upon the 1986 environmental standards to order the demolition of the Edwards Dam in Maine (completed in 1999), representing the first time that the agency had required the dismantling of a dam against the wishes of its owner.

164 **"It's impossible to moderate":** Hennessey, T. Salmon supporters take issue with feds' view of Basin Mills. *Bangor Daily News*, 14 March 1995.

164 **Baird emphasized that the real objective:** Baird, S.F. (1880). The Atlantic Salmon, pp. 28-32 in *Report of the Commissioner for 1878*, Part VI, U.S. Commission of Fish and Fisheries. Washington, DC: Government Printing Office.

164 **endangered in December 2000:** At the time of listing, the services did not have enough information to include the Penobscot River north of the Bangor dam in the distinct population segment. The 1999 Status Review found that "because potentially important and heritable adaptations are needed for larger river systems, it would be premature to determine the status of [the Penobscot River] population in relationship to the Gulf of Maine DPS without comprehensive genetic data," Colligan, M.A., J.F. Kocik, D.C. Kimball, G. Marancik, J.F. McKeon, and P.R. Nickerson (1999). *Status Review for Anadromous Atlantic Salmon in the United States*. Gloucester, MA: National Marine Fisheries Service and U.S. Fish and Wildlife Service.

165 **Maine Atlantic salmon have a higher proportion:** National Research Council (2002).

165 **In determining whether or not to list:** Kocik and Brown (2002).

165 **Federal officials invented the "wild" label:** Horvath, L. Other approaches to salmon restoration. *Bangor Daily News*, 27 December 1999.

165 **interim report on genetics:** National Research Council (2002).

166 **the determination of endangered status was sound:** *Maine v. Norton*, 257 F. Supp. 2d 357 - Dist. Court, D. Maine (2003). See also Jenkins (2003); Elmen, J. 2003. Who's afraid of the ESA: a legal analysis of listing Atlantic salmon as an endangered species. Portland, ME: University of Maine School of Law. Forest cutting, blueberry growing, aquaculture all continued after the listing.

166 **primary reason for its notoriety:** Normandeau Associates. (1995). Sears Island marine dry cargo terminal [final reports on marine resources, wildlife, wetlands, and related impacts]. Augusta, ME: Department of Transportation; Vanasse Hangen Brustlin, Inc. (1995). [draft and final environmental impact statements and related materials] Augusta, ME: Department of Transportation; Platt, R.H., and J.M. Kendra. The Sears Island saga: law in search of geography. *Economic Geography* 74:46-61; Schmitt, C. (2005). Groups envision Sears Island's future. *The Working Waterfront*, September; Schmitt, C. (2006). Sears Island: A Guide for Beginners. *The Working Waterfront*, November; Schmitt, C. (2010). Signs of stewardship on Sears Island, *The Working Waterfront*, June.

167 **finding a compromise:** Among the interests with a stake in the outcome were Protect Sears Island, Coastal Mountains Land Trust, Islesboro Island Trust, Friends of Sears Island, Penobscot Bay Alliance, Earth First!, Maine Tomorrow (led by the former commissioner of the Department of Transportation in the 1990s under Governor Angus King), Penobscot Bay River Pilots Association, Waldo County Marketing Association, Maine Pulp and Paper Association, Penobscot Bay Watch, Sierra Club, and Fair Play for Sears Island. A conservation easement now protects two-thirds of Sears Island, including nearly four miles of shoreline.

167 **John Banks knew he needed to do something:** Toensing, G.C. Dam removal launches Penobscot River restoration. *Indian Country Today*, 19 June 2012. http://indiancountrytodaymedianetwork.com/2012/06/19/dam-removal-launches-penobscot-river-restoration-118930; J. Banks, interview with the author, 30 October 2014.

167 **energy generating capacity:** Whittaker, J.A., et al. *Submittal of the Lower Penobscot River Basin Comprehensive Settlement Accord, Federal Energy Regulatory Commission with Explanatory Statement*, 25 June 2004. Washington, DC: Winston & Strawn, LLP. The dam removal idea was not new, but had been discussed during Basin Mills proceedings and proposed by multiple entities, including the Penobscot Nation.

168 **expanded the endangered listing:** Federal Register, Vol. 74, No. 117, Friday, June 19, 2009, 29344.

168 **protocols changed from those of a production facility:** Maynard, D.J., and J.G. Trial (2014). The use of hatchery technology for the conservation of Pacific and Atlantic salmon. *Reviews in Fish Biology and Fisheries* 24:803-817; Lamothe, P., and C. Domina, Craig Brook National Fish Hatchery, interviews with the author, 22 October 2014.

168 **at the expense of diversity:** Consuegra, S., and E.E. Nielsen. Population size reductions, pp. 239-269 in Verspoor et al. (2007).

168 **Fish from each river and tributary may have special characteristics:** McCormick et al. (1998).

168 **whether enough variability remains:** Hayes, S.A., and J.F. Kocik (2014). Comparative estuarine and marine migration ecology of Atlantic salmon and steelhead: blue highways and open plains. *Reviews in Fish Biology and Fisheries* 24:757-780.

168 **move in a south-south-easterly direction:** Sheehan, T.F., M.D. Renkawitz, and R.W. Brown (2011). Surface trawl survey for U.S. origin Atlantic salmon *Salmo salar*. *Journal of Fish Biology* 79:374–398.

169 **As they grow they eat more fish:** Rikardsen, A.H., and J.B. Dempson. Dietary life-support: the food and feeding of Atlantic salmon at sea, pp. 115-144 in Aas et al. (2011).

169 **back to the southern Labrador Sea:** Meister, A.L. (1984). The marine migrations of tagged Atlantic salmon (*Salmo salar* L.) of USA origin. International Council for the Exploration of the Sea CM 1984/M:27.

169 **Atlantic salmon kept declining:** Kocik and Brown (2002); Whoriskey (2009).

169 **the North Atlantic Ocean experienced a regime shift:** Mills, K.E., A.J. Pershing, T.F. Sheehan, and D. Mountain (2013). Climate and ecosystem linkages explain widespread declines in North American Atlantic salmon populations. *Global Change Biology* 19:3046-3061.
170 **large-scale climate factors influence the salmon's marine environment:** Friedland, K.D., J.P. Manning, J.S. Link, J.R. Gilbert, A.T. Gilbert, and A.F. O'Connell (2012). Variation in wind and piscivorous predator fields affecting the survival of Atlantic salmon, *Salmo salar*, in the Gulf of Maine. *Fisheries Management and Ecology* 19(1): 22-35.
170 **salmon smolts are leaving earlier:** Otero, J., et al. (2014). Basin-scale phenology and effects of climate variability on global timing of initial seaward migration of Atlantic salmon (*Salmo salar*). *Global Change Biology* 20:1-75.
170 **can tolerate warmer water:** Reddin, D.G. et al. (2012). Distribution and biological characteristics of Atlantic salmon (*Salmo salar*) at Greenland based on the analysis of historical tag recoveries. *ICES Journal of Marine Science* 69(9):1589-1597.
170 **The Atlantic salmon is *septentrional*:** Lecomte, F., E. Beall, J. Chat, P. Davaine, and P. Gaudin (2013). The complete history of salmonid introductions in the Kerguelen Islands, Southern Ocean. *Polar Biology* 36:457-475. Atlantic salmon did not succeed as well as brown trout. The authors propose that Atlantic salmon failed to navigate: "In the subantarctic islands, a magnetic compass would not be useful for Atlantic salmon to find an island in the middle of the Southern Ocean."
170 **might hold important ways to survive:** Pess, G.R., T.P. Quinn, S.R. Gephard, and R. Saunders (2014). Re-colonization of Atlantic and Pacific rivers by anadromous fishes: linkages between life history and the benefits of barrier removal. *Reviews in Fish Biology and Fisheries* 24:881-900; Kocik and Brown (2002).
170 **likely encoded in their genes:** Verspoor, E., C. Garcia de Leaniz, and P. McGinnity. Genetics and habitat management, pp. 399-424 in Verspoor et al. (2007). Kocik and Brown (2002) noted that the flexibility of wild fish could also benefit aquaculture fish, helping to address issues of productivity, disease resistance, and climatization.
171 **agencies did not include the West Branch:** The West Branch Penobscot River is currently excluded from diadromous fish species management. Dubé, N.R., R. Dill, R.C. Spencer, M.N. Simpson, O.N. Cox, P.J. Ruksznis, K.A. Dunham, and K. Gallant (2011). *Penobscot River: 2010 Annual Report*. Maine Department of Marine Resources, Bureau of Sea Run Fisheries and Habitat. The West Branch has been "off the map" of restoration since the beginning of the restoration. When Richard Cutting surveyed the Penobscot for his master's thesis in the 1950s, he concluded that restoration efforts should avoid the West Branch due to dams and pollution. Later, after it became clear that a self-sustaining population of landlocked salmon had become established and created a world-class fishery, the Department of Inland Fisheries and Wildlife prioritized the landlocked salmon, including advising against fishway construction to prevent movement of invasive species, with support from Great Northern Paper. Cutting (1956); Deason, J.P. *1993*. Letter to FERC, Maine Department of Environmental Protection files; Office of Hydropower Licensing (1996). License Renewal Application, Ripogenus and Penobscot Mills, Final Environmental Impact Statement, FERC/FEIS-0075; A. Meister, interview with the author, 22 August 2014; AuClair (2014).
171 **salmon are different:** Trinko Lake, T.R., K.R. Ravana, and R. Saunders (2012). Evaluating changes in diadromous species distributions and habitat accessibility following the Penobscot River Restoration Project. *Marine and Coastal Fisheries: Dynamics, Management, and Ecosystem Science* 4:284-293.

172 **"The culture suffers":** Phillips, B. (2006). *A River Runs Through Us*. Augusta, ME: Penobscot River Restoration Trust.

172 **Paul Greenberg argued:** Greenberg, P. (2014). *American Catch: The Fight for Our Local Seafood*. New York, NY: Penguin Press. See also Kocik and Brown (2002).

172 **rushing over stone:** Whoriskey (2009).

Selected Bibliography

Aas, Ø., A. Klemetsen, S. Einum, and J. Skurdal. 2011. *Atlantic Salmon Ecology*. Wiley-Blackwell, Ltd.

Allen, F.L. 1931. *Only Yesterday*. New York: Bantam.

Andrews, R.N.L. 2006. *Managing the Environment, Managing Ourselves: A History of American Environmental Policy,* Second Edition. New Haven, CT: Yale University Press.

Atkins, C.G. 1874. On the salmon of eastern North America, and its artificial culture, pp. 226-335 in Report of the Commissioner for 1872 and 1873, Part II, U.S. Commission of Fish and Fisheries. Washington, DC: Government Printing Office.

Atkins, C.G. 1887. The River Fisheries of Maine, pp. 673-728 in The Fisheries and Fishery Industries of the United States V (Goode, G.B., ed.), U.S. Commission of Fish and Fisheries. Washington, DC: Government Printing Office.

AuClair, S., editor. 2014. The Origin, Formation & History of Maine's Inland Fisheries Division. Rockwood, ME: Moosehead Media Services.

Babb, Cyrus C. 1912. *Certain Legal Aspects of Water-Power Development in Maine*. Augusta, ME: Maine State Water Storage Commission.

Baker, Emerson W., Edwin A. Churchill, Richard D'Abate, Kristine L. Jones, Victor A. Konrad, and Harold E.L. Prins, eds. 1994. *American Beginnings: Exploration, Culture, and Cartography in the Land of Norumbega*. Lincoln: University of Nebraska Press.

Baker, Frank L., W. Lloyd Byers, and Sam L. Worcester. *Report of the Commission to Study the Atlantic Salmon, January 1, 1947*. Augusta, ME.

Baum, E. 1997. *Maine's Atlantic Salmon: A National Treasure*. Hampden, ME: Atlantic Salmon Unlimited.

Beach, Christopher S. 1993. Conservation and Legal Politics: the Struggle for Public Water Power in Maine 1900-1923. *Maine Historical Society Quarterly* 32.3/4:150-173.

Benson, Norman G., ed. 1970. *A Century of Fisheries in North America*. Washington, DC: American Fisheries Society.

Bernier, Kevin, Kathy Billings, Norm Dube, Clem Fay, Scott Hall, Lou Horvath, and Brian Stetson. 1995. *Report of the Penobscot River Subcommittee*. Augusta, ME: Four Rivers Technical Working Group, Governor's Maine Atlantic Salmon Task Force.

Bley, P.W. 1987. Age, Growth, and Mortality of Juvenile Atlantic Salmon in Streams: A Review. Washington, DC: U.S. Fish and Wildlife Service.

Boardman, S.H. 1960. The lumber industry in the state of Maine, pp. 1224-1243 in *The New England States: Their Constitutional, Commercial, Professional, and Industrial History*, Vol. III (William T. Davis, ed.). Boston: DH Hurd & C.

Bohne, Joseph R., and Lee Sochasky, eds. 1975. *Proceedings of the New England Atlantic Salmon Restoration Conference, Boston, Massachusetts, January 14-16, 1975*, Special Publication Series No. 6. New York, NY: International Atlantic Salmon Foundation.

Bolster, W.J. 2012. *The Mortal Sea: Fishing the Atlantic in the Age of Sail*. Cambridge, MA: Harvard University Press.

Brodeur. P. 1985. *Restitution, the Land Claims of the Mashpee, Passamaquoddy, and Penobscot Indians of New England*. Boston: Northeastern University Press.

Bourque, Bruce J. 2001. *Twelve Thousand Years: American Indians in Maine*. Lincoln, NE: University of Nebraska Press.

Brewster, Ralph O. 1927. *Development and Control of Hydroelectric Power*. Augusta, ME: Maine Legislature.

Butler, J.E., and A. Taylor. 1992. *Penobscot River Renaissance: Restoring America's Premier Atlantic Salmon Fishery*. Camden, ME: The Silver Quill Press, Down East Books.

Cannon, P., and P. Brooks. 1968. *The Presidents' Cookbook*. Funk & Wagnalls.

Clay, C.H. 1961. Design of fishways and other fish facilities, including fish locks, fish elevators, fences & barrier dams. Ottawa: Canada Department of Fisheries.

Cook, David. 2007. *Above the Gravel Bar: The Native Canoe Routes of Maine*. Solon, ME: Polar Bear & Company.

Cutting, R.E. 1956. Atlantic salmon (Salmo salar) habitat in the lower West Branch, Penobscot River, Maine. M.S. thesis, University of Maine.

Cutting, R.E. 1959. Penobscot River Salmon Restoration. Orono, ME: Maine Atlantic Sea-Run Salmon Commission.

Davies, R.S. 1972. History of the Penobscot River: its use and abuse. M.S. thesis, University of Maine.

Davies, S., L. Tsomides, J.L. DiFranco, and D.L. Courtemanch. 1999. Biomonitoring Retrospective: Fifteen Year Summary for Maine Rivers and Streams, DEPLW1999-26. Augusta, ME: Department of Environmental Protection Bureau of Land and Water Quality.

Daynes, B.W., and G. Sussman. 2010. *White House Politics and the Environment*. College Station, TX: Texas A & M University Press.

Department of Commerce. 2009. Endangered and Threatened Species; Designation of Critical Habitat for Atlantic salmon (*Salmo salar*) Gulf of Maine Distinct Population Segment; Final Rule, 50 CFR Part 226. *Federal Register* 74.117 (19 June 2009):29300-29341.

Dunfield, R.W. 1985. The Atlantic Salmon in the History of North America. Canadian Special Publication of Fisheries and Aquatic Sciences 80. Ottawa: Department of Fisheries and Oceans.

Dymond, J.R. 1963. Family Salmonidae, pp. 457- 502 in Fishes of the Western North Atlantic, Sears Foundation for Marine Research Memoir Number 1, Part 3, Soft-Rayed Bony Fishes. New Haven: Yale University.

Everhart, W.H., and R.E. Cutting. 1968. *The Penobscot River Atlantic Salmon Restoration: Key to a Model River, Presented to the US Fish & Wildlife Service March 1967*. Brewer, ME: Penobscot County Conservation Association.

Fay, Clem, et al. 2006. Status Review for Anadromous Atlantic Salmon *(Salmo salar)* in the United States. Silver Spring: National Marine Fisheries Service and U.S. Fish and Wildlife Service.

Foster, N.W., and C.G. Atkins. 1868. Report of Commission on Fisheries, Senate Document No. 7, 47th Legislature. Augusta, ME.

Foster, N.W., and C.G. Atkins. 1869. Second Report of the Commissioners of Fisheries State of Maine 1868. Augusta, ME: Owen & Nash.

Foster, N.W., and C.G. Atkins. 1871. Third Report of the Commissioners of Fisheries State of Maine 1869. Augusta, ME: Owen & Nash.

Godfrey, J.E. 1882. *Annals of Bangor, History of Penobscot County Maine*. Cleveland, OH: Williams Chase & Co.

Goode, G.B. 1884. The Fisheries and Fishery Industries of the United States, U.S. Commission of Fish and Fisheries. Washington, DC: Government Printing Office.

Graham, F. Jr. 1971. *Man's Dominion: The Story of Conservation in America*. New York: M. Evans and Company.

Great Northern Paper Company. 1994. *The Hydroelectric System*. Washington, DC: Federal Energy Regulatory Commission.

Gross, M.R., R.M. Coleman, and R.M. McDowall. 1988. Aquatic productivity and the evolution of diadromous fish migration. *Science* 239:1291-1293.

Hale, R. 2005. A history of the lumber industry on the Penobscot River, Culture and History of the River Workshop, Eastern Maine Development Corporation and Penobscot River Restoration Trust, 26 January 2005, Indian Island, ME.

Hays, S.P. 1959. *Conservation and the Gospel of Efficiency*. Cambridge, MA: Harvard University Press.

Hempstead, A.G. 1931. *The Penobscot Boom and the Development of the West Branch of the Penobscot River for Log Driving*. Orono, ME: University of Maine Press.

Hendry, A.P., and S.C. Stearns, editors. 2004. *Evolution Illuminated: Salmon and Their Relatives*. Oxford University Press.

Herrington, William C., and George A. Rounsefell. 1941. Restoration of the Atlantic salmon in New England. *Journal of the American Fisheries Society* 70.1:123-127.

Holm, T. 2005. *The Great Confusion in Indian Affairs: Native Americans and Whites in the Progressive Era*. Austin, TX: University of Texas Press.

Jagels, R. 2010. Bangor pool peapods: reviving a tradition and a river. *WoodenBoat* 212:38-43.

Jarvis, N.D. 1988. Curing and canning of fishery products: a history. *Marine Fisheries Review* 50:180-185.

Jenkins, D. 2003. Atlantic salmon, endangered species, and the failure of environmental policies. *Comparative Studies in Society and History* 45(4):843-872.

Johnson, H. 1991. *Sleepwalking through History: America in the Reagan Years*. New York: Simon and Schuster.

Judd, R.W. 1997. *Common Lands, Common People: The Origins of Conservation in Northern New England*. Cambridge, MA: Harvard University Press.

Kelley, A.R., and D. Sanger. 2003. Postglacial development of the Penobscot River Valley: implications for geoarchaeology, pp. 119-133 in *Geoarchaeology of Landscapes in the Glaciated Northeast* (D.L. Cremeens and J.P. Hart, eds.) New York State Museum Bulletin 497. Albany, NY: State University of New York.

Kendall, W.C. 1935. *The Fishes of New England, The Salmon Family, Part 2—The Salmons*. Memoirs of the Boston Society of Natural History, Vol. 9, No. 1. Boston, MA: Boston Society of Natural History.

Kircheis, D., and T. Liebich. 2007. Habitat Requirements and Management Considerations for Atlantic salmon (*Salmo salar*) in the Gulf of Maine Distinct Population Segment, Draft. Silver Spring, MD: National Marine Fisheries Service.

Kocik, J.F., and R.W. Brown. 2002. From game fish to tame fish: Atlantic salmon in North America, 1798-1998, pp. 3-31 in *Sustaining North American Salmon: Perspectives Across*

Regions and Disciplines (K.D. Lynch, M.L. Jones, and W.W. Taylor, eds.) Bethesda, MD: American Fisheries Society.

Kolodny, A. 2007. A summary history of the Penobscot Nation, pp. 1-88 in *The Life and Traditions of the Red Man by Joseph Nicolar* (1893). Durham, NC: Duke University Press.

Kolodny, A. 2012. *In Search of First Contact.* Durham, NC: Duke University Press.

Kuenhnert, P. 2000. *Health Status and Needs Assessment of Native Americans in Maine, Final Report.* Augusta, ME: Department of Human Services.

Kurlansky, M. 2012. *Birdseye.* New York: Doubleday.

Lansky, M. 1992. *Beyond the Beauty Strip: Saving What's Left of Our Forests.* Gardiner, ME: Tilbury House Publishers.

MacDougall, P. 1995. *Indian Island, Maine: 1780 to 1930.* Ph.D. Diss. University of Maine.

Maine Bureau of Water Quality Control. 1971. *Interim Water Quality Plan for the Penobscot River Basin.* Augusta, ME: Department of Environmental Protection.

Maine Department of Conservation and U.S. Department of the Interior National Park Service. 1982. Maine Rivers Study: Final Report. Augusta, ME.

Maine Development Commission. 1929. *Report on Water Power Resources of the State of Maine.* Augusta, ME.

Maine Water Improvement Commission. 1963. *Penobscot River Classification Report.* Augusta, ME.

Mares, B. 1999. *Fishing with the Presidents.* Stackpole Books.

McCormick, S.D., L.P. Hansen, T.P Quinn, and R.L. Saunders. 1998. Movement, migration, and smolting of Atlantic salmon (*Salmo salar*). *Canadian Journal of Fisheries and Aquatic Sciences* 55(Suppl. 1):77-92.

McLeod, J.E. 1978. The Great Northern Paper Company. Special Collections, University of Maine Fogler Library, Orono, ME.

Milazzo, P.C. 2006. *Unlikely Environmentalists: Congress and Clean Water, 1945-1972.* Lawrence, KS: University Press of Kansas.

Mills, D. 1989. *Ecology and Management of Atlantic Salmon.* London: Chapman and Hall.

Montgomery, D.R. 2000. Coevolution of the Pacific salmon and Pacific Rim topography. *Geology* 28:1107-1110.

Nash, R.F. 2001. *Wilderness and the American Mind*, 4th ed. New Haven, CT: Yale University Press.

National Marine Fisheries Service. 2012. Endangered Species Act Biological Opinion, 29 November 2012.

National Research Council. 2002. Genetic status of Atlantic salmon in Maine: Interim Report. Washington, DC: National Academies Press.

National Research Council. 2004. *Atlantic Salmon in Maine.* Washington, DC: National Academies Press.

Netboy, A. 1968. *The Atlantic Salmon: A Vanishing Species?* Boston, MA: Houghton Mifflin.

Netboy, A. 1974. *The Salmon: Their Fight for Survival.* Boston, MA: Houghton Mifflin Company.

New England-New York Interagency Committee. 1957. Land and Water Resources of the New England-New York Region. 85th Congress Senate Document 14. Washington, DC: Government Printing Office.

New England River Basins Commission. 1981. Water, Watts, and Wilds: Hydropower and Competing Uses in New England and Penobscot River Basin Overview. Boston, MA.

Nicolar, J. 1893. *Life and Traditions of the Red Man.* Durham, NC: Duke University Press, 2007.

Osborn, W.C. 1974. The Paper Plantation: Ralph Nader's Study Group Report on the Pulp and Paper Industry in Maine. New York: Grossman Publishers.

Parrish, M.E. 1992. *Anxious Decades: America in Prosperity and Depression, 1920-1941*. New York: W. W. Norton.

Pawling, M. 2007. *Wabanaki Homeland and the New State of Maine: the 1820 Journal and Plans of Survey of Joseph Treat*. Amherst, MA: University of Massachusetts Press.

Pike, R.E. 1967. *Tall Trees, Tough Men*. New York: W.W. Norton & Company Inc.

Pratt, V.S. 1946. The Atlantic salmon in the Penobscot River. M.S. thesis, University of Maine.

Prins, H.E.L. 1994. Children of Gluskap, pp. 95-117 in *American Beginnings: Exploration, Culture, and Cartography in the Land of Norumbega* (Baker et al., eds.). Lincoln: University of Nebraska Press.

Proctor, R.W. 1942. Report on Maine Indians. Augusta, ME: Maine Legislative Research Committee.

Prysunka, A.M. 1975. The National Pollutant Discharge Elimination System and its Implementation within the Penobscot River Basin. Orono, ME: University of Maine Department of Agricultural and Resource Economics.

Reagan, R. 1990. *An American Life*. New York: Simon and Schuster.

Reiger, J.F. 2001. *American Sportsmen and the Origins of Conservation*. Corvallis: Oregon State University Press.

Rounsefell, G.A., and L.H. Bond. 1949. Salmon Restoration in Maine, Research Report No. 1, Augusta, ME: Atlantic Sea-Run Salmon Commission.

Sanger, D., and M.A.P. Renouf, editors. 2006. *The Archaic of the Far Northeast*. Orono, ME: University of Maine Press.

Sanger, D., A.R. Kelley, and H. Almquist. 2003. Geoarchaeological and cultural interpretations in the Lower Penobscot Valley, Maine, pp. 135-149 in *Geoarchaeology of Landscapes in the Glaciated Northeast* (D.L. Cremeens and J.P. Hart, eds.). Albany, NY: New York State Museum Bulletin 497.

Saunders, R., M.A. Hachey, and C.W. Fay. 2006. Maine's diadromous fish community: past, present, and implications for Atlantic salmon recovery. *Fisheries* 31(11): 537-547.

Schullery. 1999. *American Fly Fishing: A History*. New York, NY: Lyons Press.

Schurr, S.H., and B.C. Netschert; New England Energy Policy Staff. 1973. Energy in New England 1973-2000. Boston, MA: New England Reigonal Commission.

Seale, W. 1986. *The President's House: A History, Volume II*. Washington, DC: White House Historical Association and National Geographic Society and New York: Harry N. Abrams, Inc.

Shabecoff, P. 1993. *A Fierce Green Fire*. New York: Hill and Wang.

Smith, Hugh McCormick. 1898. The Salmon Fishery of Penobscot Bay and River in 1895 and 1896, pp. 112-134 in Bulletin of the U.S. Fish Commission, Volume XVII for 1897. Washington, DC: Government Printing Office.

Smith, M. 1967. *Entertaining in the White House*. Washington, DC: Metropolis Books.

Smith, David C. 1972. *A History of Lumbering in Maine, 1861-1960*. Orono, ME: University of Maine Press.

Smith, A.F. 2004. *The Oxford Encyclopedia of Food and Drink in America*. New York: Oxford University Press.

Smythe, O.M. 1953. Historical account of fishing activity at famous Bangor Salmon Pool. *Bangor Daily News* (18 May 1953):14.

Soden, D.L., ed. 1999. *The Environmental Presidency*. Albany, NY: State University of New York Press.

Speck, Frank G. 1997. *Penobscot Man: the Life History of a Forest Tribe in Maine.* Orono, ME: University of Maine Press.

Stearley, R.F. 1992. Historical ecology of Salmoninae, with special reference to Oncorhyncus, pp. 622-658 in *Systematics, Historical Ecology, and North American Freshwater Fishes* (R.L. Mayden, editor). Stanford, CA: Stanford University Press.

Stearley, R.F., and G.R. Smith. 1993. Phylogeny of the Pacific trouts and salmons (*Oncorhynchus*) and genera of the family Salmonidae. *American Fisheries Society Transactions* 122:1-33.

Stevens, D.H. 1952. Report to Legislative Research Committee Regarding Indian Affairs. Augusta, Maine Department of Health and Welfare.

Swain, D.C. 1963. *Federal Conservation Policy, 1921-1933.* Berkeley, CA: University of California Press.

Thorstad, E.B., F. Whoriskey, A.H. Rikardsen, and K. Aarestrup. Aquatic nomads: the life and migrations of the Atlantic salmon, pp. 1-32 in Aas et al. (2011).

Thoreau, H.D. 1864. *The Maine Woods.* New York, NY: Harper and Row, 1987.

USASAC (U.S. Atlantic Salmon Assessment Committee). Annual Report. Turners Falls, MA: NASCO, multiple years.

US EPA. 1980. A Water Quality Success Story: Penobscot River, Maine. Washington, DC: Office of Water Regulation and Standards.

Verspoor, E., L. Stradmeyer, and J.L. Nielsen. 2007. *The Atlantic Salmon: Genetics Conservation and Management.* Oxford, UK: Blackwell Publishing.

Walton, I. 1939. *The Compleat Angler.* New York: The Modern Library.

Wells, H.P. 1886. *The American Salmon Fisherman.* London: S. Low, Marston, Searle & Rivington.

Wells, W. 1868. The Water-Power of Maine: Reports of the Commissioners and Secretary of the Hydrographic Survey of 1867. Augusta, ME: Owen & Nash.

Whitcomb, J., and C. Whitcomb. 2000. *Real Life at the White House.* New York, NY: Routledge.

Whoriskey, F. 2009. Management of angels and demons in the conservation of the Atlantic salmon in North America. *American Fisheries Society Symposium* 70:1-19.

Wilkins, A.H. 1932. The Forests of Maine, Bulletin No. 8. Augusta, ME: Maine Forest Service.

Index

Adams, Abigail, 20
Adams, John, 20
Alamoosook Lake, 141, 142
Alaska salmon, 53–54, 55
alevins, 71
alewives, 1, 13, 162
Allagash River, 120
Allen, Frederick, 52
Allen, Ida, 80–81
Aluminum Company of America, 25
American Angler's Book, 30
American Catch, 172
American Revolution, 37, 176n; salmon sustaining troops, 20
American Rivers, 137, 154, 158
The American Salmon Fisherman, 21
Anadromous Fish Conservation Act, 120
Anderson, Karl, 27, 28, 51; call for regulations, 75; first-fish ritual, 20–21, 22, 24; peapod boat, 19, 22, 29
Androscoggin River, 1, 2, 101, 107, 108
Appalachian Mountain Club, 137, 204n
aquaculture, 149–150, 157
Army Corps of Engineers, 25, 59, 99; on dams, 112; emergence, 187n; permits, 99; reviewing status of salmon, 158
Arnold, Benedict, 20
Arthur, Chester, 4
Arts and Crafts movement, 28
Aspegeunt, Joseph, 39

Atkins, Charles, 34, 70, 71, 72, 73; egg production and, 73, 75, 142; on mill dust, 99, 100; on spawning habits, 143; on stocking, 84
Atlantic salmon, 1; adapting over time, 16; aquaculture, 149–150, 157; canning, 53; climate change and, 169–170; as endangered species, 6, 110, 157, 159–160, 165–166, 171; Fish and Wildlife Service on, 157; fly-fishing, 21; genetically distinct, 165; Greenland fishery, 145–146, 147; hatcheries preventing extinction, 168; imprinting, 35; King of Fish, 1, 58, 77, 82, 161; luring, 29; migration patterns, 34–35; name derivation, 23–24; Pacific salmon compared to, 17; Penobscot River, 20, 21–22, 53, 53–54, 94; Reagan, R., and, 139; restoration of, 2–3, 86–87, 110–111; scarcity, 149; waterflow and, 94
Atlantic Salmon Federation, 131
Atlantic Sea-Run Salmon Commission, 87–88, 103, 105, 109; on fishways, 120; river restoration plan, 110; stocking by, 124
Atlantic States Marine Fisheries Commission, 86
Ayer, Fred, 21
Aylward, David, 84

Baird, Spencer F., 70, 163, 164
Baldacci, John, 167
Ballinger, Richard, 25
Bangor Daily Commercial, 27
Bangor Daily News, 48, 80, 131, 142, 151, 154
Bangor Dam, 82, 84; break in, 122; building original, 128; defeat of, 152; feasibility study, 135; lawsuit, 140; license denial, 138; opposition, 131, 133, 135–136; rebuilding, 128, 130
Bangor Electric Light and Power Company, 26
Banks, John, 148, 167
barracuda, 4
Basin Mills Dam, 41, 152–156, 157, 158–159, 164
bass, 1, 13, 94, 155, 163
Baum, Ed, 151–152, 156, 157
Baxter, Percival, 48–49
Bayly, Julia, 149
beauty strips, 34
beaver, 11, 13, 176n
Beothuks of Newfoundland, 5
Big Ambejackmockamus Falls dam proposal, 136–138, 205n
Big Eddy Dam, 43
Big Niagara Falls, 36
Bingham, William, 37
Birdseye, Clarence, 58
Bissell, Charles, 51
Bisulca, Paul, 154
Bond, Horace, 79–80, 84, 109
Bond, Lyndon, 87
bonefish, 4
Bonneville Dam, 81, 83
Boston Cooking-School Book, 20
Bowden, Avery, 163
Brennan, Joseph, 127
Bruckner, Katherine, 59
Bryant, Nelson, 154
Bureau of Fisheries: hatchery program, 55; Hoover, H., and, 54, 55; product laboratory, 54
Bush, Barbara, 3
Bush, George H. W., 127; fishing president, 4; National Energy Strategy, 6; Presidential Salmon to, 2–4, 6, 7, 152, 156

Bush, Henderson P., 3
Butterfield, Joseph, 68

Canning, J. Edward, 56
carbaryl, 119
Caron, Charles, 140
carp, 73
Carr, Elisha, 142
Carroll, Guy, 89
Carson, Rachel, 86, 106
Carter, Jimmy, 125, 127; environmental policies, 134; National Energy Act of 1978 and, 130
Cascapedia River, 4
Caucomgomoc Lake, 43
Chadwick, Joseph, 39
Chapman, Horace, 59
Cheney, A. N., 143
Cheney, S. A., 59–60
Chesuncook Lake, 120, 133; calm expanse, 36; Chadwick on, 39; dams, 39, 41, 43, 44–45; Great Northern Paper Company raising, 49; hydroelectric power, 43; landlocked salmon, 34; logging, 38; mapping, 39–41; Penobscots on, 39; routes to, 40; salmon, 43; Thoreau visiting, 42, 45; Treat on, 40–41; unorganized territory, 33
chlor-alkali process, 113
chub, 94
Church, Frederic, 63
Churchill, Winston, 80, 89
Civilian Conservation Corps, 81
Clean Water Act, 118
Cleveland, Grover, 4
climate change, 169–170
Clinton, Bill, 157
Coast Guard, 4, 115
cod, 145, 163
Cohen, William S., 127
Colby, Timothy, 116
Colorado River Storage Project, 90
Columbia River, 53, 139; Coulee Dam, 59, 83
commercial fishing, 88, 145; Canada, 146; decline, 28; hatcheries and, 55, 73, 74; managing, 86; meccas for, 111; Meister and, 147; NASCO ending, 169; regulations, 75, 82; stopping, 29, 86.

See also recreational fishing
Connecticut River, 16, 43
Conservation Law Foundation, 154
Convention for the Conservation of Salmon in the North Atlantic Ocean, 146
Coolidge, Calvin, 52; conservation, 56–57; fishing, 58; presidency, 56
Creager, Gus, 52
Crie, Horatio D., 60
Crowell, John, 135
cunner, 163
Curtis, Kenneth, 129
cusk, 94
Cutting, Richard, 100, 111, 117, 144

dace, 94
dams: Army Corps of Engineers on, 112; Basin Mills Dam, 41, 152–156, 157, 158–159, 164; Big Ambejackmockamus Falls dam proposal, 136–138; Big Eddy Dam, 43; Bonneville Dam, 81, 83; Chesuncook Lake, 39, 41, 44–45; Coulee Dam, 59, 83; Dolby Dam, 101; General Dam Act of 1906, 25; Grand Coulee Dam, 83; Great Northern Paper Company on, 120; Great Works Dam, 41, 43, 84, 112, 154; Hetch Hetchy Dam, Yosemite National Park, 49; Hoover, H., and, 55; Hoover Dam, 81; licensing, 112; Mattaceunk Dam, 83, 86; Medway Dam, 93; Milford Dam, 99; North Twin Dam, 91; obstructing waterflow, 94; Penobscot River, 25–27, 42, 74–75; Penobscot River proposal, 6, 7; public good decisions, 111; Quakish Dam, 101; Ripogenus Dam, 43, 44, 49, 60, 91; roll, 37; salmon at, 43; sawmills and, 41; Slide Dam, 43; splash, 37; Taft, W. H., and, 25; Veazie Dam, 26, 41, 75, 82, 122, 154; Westfall, C., and, 6; Wilson, W., approving, 49–50. *See also* Bangor Dam; hydroelectric power; pulp mills; sawmills
Davis, Harry, 105
DDT, 119
D'Errico, Roger, 125, 130
Diamond, John, 153

Dillingham, Mrs. George W., 21
dioxins, 148, 208n
Dodge, David, 142
Dolby Dam, 101
Donnelly, Neil, 139
Douglas, William O., 111
Dow, Raymond, 82
dry-kill, 39
Duchamp, Marcel, 28
Dymond, John, 21
dynamite, 38, 74

eagles, 37, 96, 137, 143, 163
Eckstorm, Fannie Hardy, 43
Eddington Salmon Club, 153, 159
Edging Law, 99
Edwards, James, 135
eels, 1, 64, 162
Eisenhower, Dwight: Colorado River Storage Project, 90; hydroelectric power and, 89; on natural resources, 194n; Presidential Salmon to, 89, 92, 103; wastewater treatment plants and, 107
Eisenhower, Mamie, 89
Electric Consumers Protection Act of 1986, 138–139
electricity, 41; consumption, 112; deregulating, 6; exporting, 48; rivers and, 26
Elliot, Jeffery, 157
Ellison, Bill, 151–152
Emerson, Hawley, 14
Endangered Species Act, 6–7, 110, 138
England, first-fish ritual, 23
Environmental Defense Fund, 121
Environmental Protection Agency (EPA), 118, 120, 134–135, 135; dissolved oxygen criteria, 137
erosion, 11, 88
Everhart, Harry, 111
Evert, Chris, 4
extinction: hatcheries preventing, 168; Pacific salmon threat, 149; salmon, 70

Fawcett, James, 122
Federal Aid in Sport Fish Restoration Program, 121

Federal Energy Regulatory Commission (FERC), 130, 138–139, 167
Federal Power Act, 50, 112
Federal Power Commission, 81, 121
Federal Water Pollution Control Act, 118
Fellows, Frank, 85
FERC. *See* Federal Energy Regulatory Commission
Fernald Law, 48, 184n
Fickett, Oscar, 24
Fiddler Island, 96
fingerlings, 72
first-fish ritual, 171–172; Anderson and, 20–21, 22, 24; Bond and, 79–80; Canning and, 56; Carroll and, 89; Davis and, 105; Ellison and, 151–152; England, 23; Flanagan and, 51; honoring sea, 23; Japan, 23; Mallett and, 125; Presidential Salmon as, 24; renewal of, 106; Smith, D., and, 85; Westfall S., and, 156–157. *See also* Presidential Salmon
Fischer, Albert, 51
Fish and Wildlife Coordination Act, 121, 138
Fish and Wildlife Service, 6, 84, 85, 86; agreements, 87; on Atlantic salmon, 157; dam licensing, 153; hatcheries, 121; restoration plan, 110
Fish Commission, 70, 73
Fish Conservation Act, 111
fishing: by Bush, G. W., 4; by Coolidge, 58; Hoover, H., on, 55–56; ice fishing, 30; laws, 69; limits, 29, 30; during migration, 67; overfishing, 70, 84; by Penobscots, 13, 29–30; pound-net, 14–15; rights of Penobscots, 67–69, 147, 154–155, 172; sawmills and, 70–71; spear, 30–31; treaties, 88; by Washington, 4. *See also* commercial fishing; first-fish ritual; fly-fishing; recreational fishing
fish ladders, 28, 55, 83, 91, 173n; conveyance by, 88, 91; installing, 55; negotiating, 83; Penobscot River, 60; placement, 87
fishways, 82–84, 111; Atlantic Sea-Run Salmon Commission on, 120; not working, 86; Penobscot River, 120–121

Flanagan, Michael: first-fish ritual and, 51; Presidential Salmon and, 52
flashboards, 84
Fletcher, George, 139
Flood Control Act, 90
flounder, 163
Floyd, Joe, 131
fly-fishing, 21; as gentle art, 28–29
Fly Rod & Reel, 4
Fonda, Henry, 107
Food Administration, 50–51
Ford, Gerald, 124
Forest & Stream, 28, 143
Foster, Nathan W., 70, 99
Franz Ferdinand, 48
Freese Island, 96
freezing fish, 57–58
French, Edward, 159
Friends of the Penobscot River, 131, 137
fry, 71–72, 73

Garfield, James, 25
Geagan, Bill, 80
General Dam Act of 1906, 25
General Electric, 25, 130, 202n
Geological Survey, U.S., 27
Gorbachev, Mikhail, 139
Gorbachev, Raisa, 139
Gore, Al, 157
Gorsuch, Anne, 134
Gould, E. W., 51
Grand Coulee Dam, 83
Grant, Ulysses S., 70
Grass Island, 96
grayling, 15, 73
Great Depression, 59
Great Naturalist, 133
Great Northern Paper Company, 57; Big Ambejackmockamus Falls dam proposal, 136–138, 205n; on dams, 120; hydroelectric power, 6, 91; as octopus, 48; pulp mills, 44, 109; raising Chesuncook Lake levels, 49; Ripogenus Dam, 44, 49, 60, 91; sawmill, 44; wastewater treatment plant, 119; West Branch Driving and Reservoir Dam Company, 49
Great Works Dam, 41, 43, 84, 112, 154
Greenberg, Paul, 172

Index

Greenland fishery, 145–146, 147
Greenleaf, Moses, 40
Griffith, Richard E., 102
Grinnell, George Bird, 28

hake, 163
Hale, Frederick, 52
Hamlin, Hannibal, 26
Harding, Florence, 52
Harding, Warren G.: accessibility, 52; death of, 56; economy and, 51; Presidential Salmon to, 51–52; resource development, 52
Hardy, Manly, 42, 43, 64
Haskell, Robert, 129
hatcheries: Alamoosook Lake, 142; artificial propagation, 71; Bureau of Fisheries program, 55; commercial fishing and, 55, 73, 74; Fish and Wildlife Service, 121; Penobscot River program, 55, 71, 73, 110, 121–122; preventing extinction, 168; profit from, 74; saving eggs, 72–73; supplementing salmon, 2
Hennessy, Tom, 164
herring, 1, 4, 162, 163
Herrington, William, 85
Herter, Christian, 137
Hetch Hetchy Dam, Yosemite National Park, 49
Hildreth, Horace, 86
Hog Island, 97
Holmes, Ezekiel, 64
Hoover, Herbert: Bureau of Fisheries and, 54, 55; dams and, 55; on fishing, 55–56; Food Administration head, 50–51; grandeur, 58; natural resources and, 61; Presidential Salmon to, 59
Hoover, Lou, 58
Hoover Dam, 81
hornpout, 94
Howard, Alice, 24
Hubbard, Lucius, 43
Hudson River, 16, 121
Hunt, William, 63
Hunt Farm, 63–64, 64, 66–67, 74
Hussein, Saddam, 5
hydroelectric power, 6, 43, 81; Chesuncook Lake, 43; early, 26;
Eisenhower, D., and, 89; government control, 49; Great Northern Paper Company, 6, 91; legal challenges, 121; licensing dams, 112; outweighing adverse impacts, 132; Penobscot River, 27, 90–91, 112, 129–130; private projects, 50; rise in, 48; Taft, W. H., and, 25. *See also* dams; electricity

ice fishing, 30
Ickes, Harold, 83
imprinting, 35, 102, 143, 181n
Indian Claims Commission Act, 124
International Minerals and Chemicals, 113, 198n

Jaffray, Elizabeth, 24, 47
Japan, first-fish ritual, 23
Johnson, Lyndon B.: Clean Water by 1975 pledge, 108; Presidential Salmon to, 105–106; vision, 111; Wild and Scenic Rivers Act, 111; Wilderness Act, 106–107
Judd, Richard, 70

kelts, 65, 126, 162
Kendall, William Converse, 77, 101, 144
Kenduskeag Stream: canoe race, 117, 120; cement narrowing, 117; ice melt, 115–116; origins and flow, 116; pollution, 117; salmon assessment, 117; salmon in, 116; salmon returning to, 123–124; winter tide, 115
Kennebec River, 1, 2, 43, 101, 164
Kennebunk River, 2
Kennedy, John F., 103, 106; water quality and, 107
Ketchum, Louis, 64
King, Angus, 159
King, William, 40, 67
king of fish, 1, 58, 77, 82, 161
King Pines, 37, 181n
Ku Klux Klan (KKK), 52

lamprey, 1, 162, 163
Land and Water Resources Council, 132
landlocked salmon: in Chesuncook Lake, 34; Sebec Lake, 94; spreading, 73
Lane, Franklin, 50

LaRock, John, 58
leapers, 36, 94
Leavitt, Bud, 142, 154
Leavitt, J. F., 21
Leonard, Hiram, 21, 29
Leopold, Aldo, 88
Linnaeus, Carl, 23
Lockwood, A. D., 26
logging: Chesuncook Lake, 38; manipulating river flows, 75–77; Penobscot River, 37–38, 41–42, 75; Wassataquoik Stream, 63. *See also* sawmills
Longest Walk protest, 125
Love Canal, 134
Lujan, Manuel, 7
lumpfish, 163

mackerel, 4
Magnuson Stevens Act, 146
Maine Folklife Center, 163
Maine Indian Land Claims Settlement Acts, 125
Maine Rivers Act, 133, 135, 138
Maine Rivers Study, 132, 204n
Mallett, Ivan, 125–126, 127
Mallett, Phyllis, 127
manufacturing, 57
Marancik, Jerry, 156
Mattaseunk Dam, 83, 86
Mattawamkeag River, 41, 44, 72, 93–94
McCann, Paul, 120
McIntyre, Clifford, 105
McKernan, John, 3, 140
McNally, John, 105
Medway Dam, 93
Meister, Al, 120, 144; commercial fishing and, 147
menhaden, 163
mercury, 113, 148, 198n
Merrimack River, 16, 43
migration: Atlantic salmon patterns, 34–35; fishing during, 67; to Penobscot River, 2, 35–36; smolt out-migration, 162. *See also* spawning
Migratory Bird Treaty, 55
Milardo, Bob, 153
Milford Dam, 99
Mill Act of 1821, 26

Miller, Charley, 85
Millinocket Lake, 36, 43, 44
Mink Island, 97
minnows, 94
Mitchell, George, 125, 139
Mitterrand, Francois, 4
Moody, Augustus, 119
Moore, Elizabeth, 80–81
Mowat, Farley, 145
Muir, John, 49
Muskie, Edmund, 107–108, 109, 110; clean water and, 118; as Secretary of State, 125

NASCO. *See* North Atlantic Conservation Organization
National Conservation Commission, 25
National Energy Act of 1978, 130
National Environmental Policy Act, 121, 138
National Park Service, 50
National Pollutant Discharge Elimination System, 119
National Wildlife Federation, 84
Natural Resources Defense Council, 121
Nelligan, Patrick, 47
Neptune, William, 67
Nesbitt, Henrietta, 80, 85
netting, 4, 14, 74; gillnetting, 88, 144, 147; pound-net fishing, 14–15
New Deal, 81, 97, 118
New England River Basins Commission, 132, 202n
Nicola, Albert, 98
Nixon, Richard, 124; Clean Water Act vetoed by, 118
Non-Intercourse Act of 1790, 124
North Atlantic Conservation Organization (NASCO), 146, 169
Northern spotted owl, 6, 7
North Twin Dam, 91
Norway, 4, 145, 149

oil embargo, 129, 136
Olamon Island, 96
Orson Island, 98
osprey, 37, 143, 166
over-civilization, 28
overfishing, 70, 84

Index

Pacific salmon, 17, 65, 82; canning, 53–55; Columbia River, 53; extinction threat, 149
Packard, Alpheus, 64
paper, 13, 48. *See also* Great Northern Paper Company; pulp mills
parr, 72; smolting of, 115
Passamaquoddy Tribe, 12, 124, 125
Payne, Frederick, 89
peapod boats, 19, 22, 29, 116
Pennsylvania Power and Light of Maine, LLC (PPL), 167
Penobscot Coalition, 153–154, 155
Penobscot Log-Driving Company, 41
Penobscot River, 1; ancient travel route, 11–12; Atlantic salmon, 21–22, 53, 53–54, 94; banks of, 2; branches meeting, 93–94; commercialization, 66; dam proposal, 6, 7; dams, 25–27, 42, 74–75; decline in salmon, 28; dwindling salmon, 67; as endangered, 154; first salmon in, 10–11, 12; fish ladders, 60; fish varieties, 162–163; fishways, 120–121; fly-fishing on, 21; formation, 9–10, 174n; Friends of the Penobscot River, 131, 137; hatchery program and, 55, 71, 110, 121–122; hydroelectric power, 27, 90–91, 112, 129–130; islands, 96–98, 97, 98, 161, 166–167; ledges, 98–99; logging, 37–38, 41–42, 75; mercury in, 111, 113; in 1912, 19–20, 28; origination, 16; Penobscots on, 12–13; pollution, 99–102, 103–104, 108–110, 118; power sites, 59–60, 187n; Presidential Salmon from, 5, 7; protecting salmon, 60; pulp mills, 43, 109; pulp mill waste, 100–102, 118–119; rank odor, 102; reservoirs, 95; restocking, 73; restoration plan, 111, 171; Rhine of Maine, 141; salmon migrating, 2, 35–36; salmon returning, 113; sawmills on, 37, 181n; smolts, 95–96; spawning grounds, 63, 122; spring flood, 122, 200n; strangers on, 12–13, 14, 15; waste in, 99–100; waterflow, 94–95; watershed, 9, 33, 36, 182n; Wild and Scenic Rivers system, 111, 120; during World War I, 50. *See also* Alamoosook Lake; Chesuncook Lake; Kenduskeag Stream; Seboomook Lake; Wassataquoik Stream
Penobscot River Renaissance, 156
Penobscots, 116; ancestors, 13–14; blocked from spear fishing, 30–31; on Chesuncook Lake, 39; crop subsidies, 180n; displacement of, 37; fishing rights, 67–69, 147, 154–155, 172; fishing salmon, 13, 29–30; as guides, 39; pursuing restitution, 124–125; river islands and, 96–98; threatened by settlers, 15; Treat and, 40; treaties, 13, 176n; water resources program, 148–149, 208n
Penobscot Salmon Club, 21, 28, 29, 31, 82; membership, 123; rules, 123; salmon restoration and, 84, 91–92
perch, 94
Phillips, Butch, 172
philopatry, 35
pickerel, 64, 94
Pinchot, Gifford, 25, 27, 48
pines, 37, 42
Piscataquis River, 94
Platt, David, 133
Pliny the Elder, 5, 23
plummets, 11, 175n
pollock, 143, 144, 163
pollution. *See* water pollution
Poor, John, 26
pound-net fishing, 14–15
PPL. *See* Pennsylvania Power and Light of Maine, LLC
Presidential Salmon: as arduous, 157; to Bush, G. W., 2–4, 6, 7, 152, 156; Chapman and, 59; to Eisenhower, D., 89, 92, 103; end of tradition, 92, 112; as first-fish ritual, 24; Flanagan and, 52; Fletcher and, 139; to Harding, W. G., 51–52; to Hoover, H., 59; to Johnson, 105–106; King of Fish and, 58; from Penobscot River, 5, 7; to Reagan, R., 126, 127, 139, 140; to Roosevelt, F. D., 80–81; Sullivan catching, 47; to Taft, W. H., 24–25, 32, 88, 117; to Truman, H., 85, 86; to Wilson, 47
Prohibition, 52

Public Utilities Regulatory Polices Act (PURPA), 130
pulp mills, 43; Great Northern Paper Company, 44, 109; Penobscot River, 43, 109; rise in, 48; waste in Penobscot River, 100–101, 118–119
PURPA. *See* Public Utilities Regulatory Polices Act

Quakish Dam, 101
Quakish Lake, 44
Quoddy Tides, 159

Rabeni, Charles, 118
Reagan, Nancy, 128
Reagan, Ronald, 4, 125; Atlantic salmon and, 139; dining habits, 127–128; environmental policies, 134–135, 138; General Electric and, 130, 202n; as Great Naturalist, 133; Presidential Salmon to, 126, 127, 139, 140; state dinners, 139, 205n
recreational fishing, 58; Canada, 146; rules, 87; shutting down, 92
redds, 72; changing waterflow and, 76
Red Scare, 52
Reed, John H., 108
Refuse Act, 99
Restigouche River, 5
Rhine of Maine, 141
Ripogenus Dam, 43, 44, 49, 60, 91
Ripogenus Lake, 36, 40
Rivers and Harbors Act, 59
roll dams, 37
Roosevelt, Eleanor, 80
Roosevelt, Franklin D., 61; on fishways, 83; New Deal, 81, 97, 118; Presidential Salmon to, 80–81; water pollution and, 102
Roosevelt, Theodore, 24–25, 31, 47; progressive ideology, 48
Rounsefell, George, 85, 87
Ruhlin, Dick, 113, 131, 156
runarounds, 37

Saco River, 1, 2, 43, 101
sailfish, 4
salmon: Alaska salmon, 53–54, 55; ancestry, 15–17; Army Corps of Engineers reviewing status, 158; Chesuncook Lake, 43; Chinook, 73; coho, 73, 86; compared to tuna, 86; counting, 2, 173n; at dams, 43; decline in Penobscot River, 28; as delicacy, 20; extinction, 70; first in Penobscot River, 10–11, 12; fisheries, 23; fishing laws, 69; fishing limits, 29, 30; getting smaller, 28, 179n; as immortal, 22; in Kenduskeag Stream, 116, 117, 123–124; life-cycle, 168–169; Maine, 3; migratory fish and, 164; 1912 harvest, 19; once thriving, 1, 2; in Penobscot River, 2, 5, 7, 20; Penobscots fishing, 13, 29–30; philopatry, 35; pound-net fishing, 14–15; preparing and eating, 124; protecting, 60; purity of stock, 164, 165; reddish flesh, 5, 174n; as royal, 5, 14; in Seboomook Lake, 10; served during World War I, 50–51; smell and, 35, 102, 181n; smoking, 15; sockeye, 86; sport catch, 111; stocking, 73–74; studying and tagging, 143–145; supplemented in hatcheries, 3; sustaining Revolutionary War troops, 20; throughout history, 4–5; weirs for trapping, 14–15; wild, 164–165. *See also* Atlantic salmon; fingerlings; fry; Pacific salmon; parr; Presidential Salmon; redds; smolts; *specific topics*
Salmon City, USA, 131, 135
Salmo salar, 16, 23–24, 72, 123, 157, 161, 170
A Sand County Almanac (Leopold), 88
sawmills: dams and, 41; fishing and, 70–71; fishing laws and, 69; Great Northern Paper Company, 44; on Penobscot River, 37, 181n; shipping of products, 42; transport to, 20
Schenck, Garrett, 49
Scott, Matt, 137
sculpin, 163
Sea of Slaughter (Mowat), 145
Sears Island, 161, 166–167
Sebec Lake, 94
Seboomook Lake, 43, 133; ancient activity on, 11; history, 11; salmon arriving, 10
Sewall, Joseph, 131

shad, 1, 4, 13, 47, 73, 162, 163
Shields, Don, 153
shiners, 94
Sierra Club, 121, 154, 166
Silent Spring (Carson), 106
Slide Dam, 43
sluiceways, 38
smelt, 1, 162, 163
Smith, Donald, 85
Smith, Hugh M., 54, 55, 73, 144, 161
smolts, 95–96, 121; imprinting by, 143; out-migration, 162; travel speed, 143; vulnerability, 143
Snake River, 111, 112
Snowe, Olympia, 3, 159
spawning: Atkins on, 143; Penobscot River, 63, 122; process, 65–66; Wassataquoik Stream, 64–65, 77
splash dams, 37
spruce, 43
spruce budworm, 119, 199n
Standard Oil Company, 44
steelhead, 4, 73
Stevens, David, 97
sticklebacks, 94
Stillwell, E. M., 74
stocking: Atkins on, 84; by Atlantic Sea-Run Salmon Commission, 124; Penobscot River restocking, 73; procedures, 85; salmon, 73–74
Stone, Livingston, 31
sturgeon, 1, 4, 13, 162, 163
Sugar Island, 96
Sullivan, Jeanette, 51, 56; catching Presidential Salmon, 47
Summer Vacations at Moosehead Lake and Vicinity (Hubbard), 43
sunfish, 94
Sweden, 4

Taft, Nellie, 24
Taft, Peter, 125
Taft, William Howard, 47; dam construction and, 25; hydroelectric power and, 25; Presidential Salmon to, 24–25, 32, 88, 117; weight, 31
tarpon, 4
tautog, 163
Tennessee Valley Authority, 81

Thomas, Fred, 29
Thomas Rod Company, 29
Thoreau, Henry David: on Bangor, 116, 198n; at Chesuncook Lake, 42, 45; at Wassataquoik Stream, 63–64
Three Mile Island nuclear power disaster, 127
timber cruisers, 37, 182n
tomcod, 1, 116, 162
Townsend, Bill, 120, 156
Treat, Joseph, 40–41, 42
Troubled Waters, 107–108
trout, 43, 94
Truman, Harry: Colorado River Storage Project, 90; fishing treaties, 88; Flood Control Act, 90; Presidential Salmon to, 85, 86; water pollution and, 102
tuna, 86

Udall, Stuart, 110

Veazie, Samuel, 99
Veazie Dam, 26, 41, 75, 82, 122, 154
Veazie Salmon Club, 2, 3, 123, 154, 159
Verdon, Rene, 106
Vikings, 5
vitamins, 57
Viviani, M. Rene, 52

Wabanaki Confederacy of Native American Tribes, 12, 40, 176n; unification, 98. *See also* Penobscots
Waldo, Samuel, 161
Walker, James, 152
Walton, Isaak, 5, 34, 143
Wardwell, Butch, 125
Warren, Joseph, 101
Washington, George: fishing president, 4; natives and, 13
Washington Post, 131
Wassataquoik Stream: logging, 63; spawning grounds, 64–65, 77; Thoreau at, 63–64
Wassumkeag Island, 161
wastewater treatment plants, 107
water pollution, 70, 107; Federal Water Pollution Control Act, 118; Kenduskeag Stream, 117; National Pollutant Discharge Elimination

System, 119; Penobscot River, 99–102, 103–104, 108–110, 118; Roosevelt, Franklin D. and, 102; Truman and, 102
Water Quality Act of 1965, 109
Watt, James, 135
weirs, 14–15
Wells, Henry P., 21
West Branch Driving and Reservoir Dam Company, 49
West Enfield Dam, 94
Westfall, Claude, 1, 156; delivering Presidential Salmon, 2–4, 6, 7, 156; experienced fisherman, 122–123; Penobscot River dam and, 6; Veazie Salmon Club founding, 123, 154
Westfall, Rosemae, 1, 2, 3, 4, 7
Westfall, Scott, 156–157
Westinghouse, 25
whitefish, 94
white sucker, 94
Wild and Scenic Rivers Act, 111–112, 120
Wilderness Act, 106–107
Williams, Ted, 122, 154, 161
Wilson, Woodrow: dam approval bills, 49–50; election win, 47; failing health, 51; Federal Power Act and, 50; Hetch Tetchy Dam bill, 49–50; National Park Service creation, 50; Presidential Salmon to, 47
Works Progress Administration, 82
World War I, 48; natural resources during, 53; Penobscot River during, 50; salmon served during, 50–51
World War II, 85; eating during, 86
Wrengzyk, Bob, 153
Wright, Zephyr, 106